Journalism and Technological Change

Clemens Zimmermann is Professor of Cultural- and Media History in Saarbrücken. *Martin Schreiber* works as an Official Advisor for Quality Management at Trier University.

Martin Schreiber, Clemens Zimmermann (eds.)

Journalism and Technological Change

Historical Perspectives, Contemporary Trends

Campus Verlag
Frankfurt/New York

Bibliographic Information published by the Deutsche Nationalbibliothek.
The Deutsche Nationalbibliothek lists this publication in the Deutsche Nationalbibliografie; detailed bibliographic data are available in the Internet at http://dnb.d-nb.de
ISBN 978-3-593-50104-8

Cover design: Campus Verlag, Frankfurt-on-Main
Cover illustration: on the left: editorial office of the Ullstein-Haus (at the end of the 1920s) © W. Joachim Freyburg, Hans Wallenberg (Hrsg.): Hundert Jahre Ullstein. Band 1, Berlin 1977, S. 177. On the right: Newsroom of the German daily newspaper "Die Welt" (2010) © Axel Springer SE, Berlin
Typeset: Marion Gräf-Jordan, Heusenstamm
Printing office and bookbinder: CPI buchbuecher.de, Birkach
Printed on acid free paper.
Printed in Germany

www.campus.de
www.press.uchicago.edu

Contents

Preface

This volume is based on a conference that took place on 22.–23. März 2013 at Villa Lessing in Saarbrücken, organized by the Chair for Cultural and Media History at *Saarland University*. The conference brought together international experts in the research field of media and journalism studies from different disciplines and countries – and we hope that this volume reflects some of its inspiring atmosphere which was made possible by many supporters:

First of all, the generous financial aid of *VolkswagenStiftung*, and particularly the assistance of the responsible consultant Wolfgang Levermann, made the realization of the conference and the production of this volume possible. For the planning and organization of the conference, the Saarbrücken staff – Susanne Dengel, Aline Maldener, Michael Röhrig and Heike Werner – once more proved invaluable. We thank all of them very much for their great commitment. Further thanks should be given to Volker Linneweber, the President of *Saarland University*, and to Michael Kuderna from *Saarländischer Rundfunk* for their illustrative and revealing introductory contributions. Of great importance and indispensable was the work undertaken by Ilka Braun and, especially, Aline Maldener at the Chair of Cultural and Media History who edited the manuscripts with great care and indefatigable attention.

We would also like to thank our editor Stefanie Evita Schaefer from the publisher Campus Frankfurt/New York for her practical support and her many valuable ideas. Judith Thissen (Utrecht) helped us with a creative proposal regarding the title of the book. Special thanks are also due to the committed translation and editorial works of Anna Richter (Berlin), Rebecca van Dyck (Hannover), Alex J. Kay (Frankfurt), Janice I. Horton and Nikola Langreiter from *Wortstellerei* (Lustenau, Austria).

Saarbrücken, May 2014 — *Martin Schreiber, Clemens Zimmermann*

Introduction: Towards a New Perspective on Journalism and Technology

Clemens Zimmermann and Martin Schreiber

Technologies, media and journalism are closely interrelated. This is true for the present time as much as in historical perspective. Technologies such as telegraphy and the rotation press contributed to the multiplication and globalization of news and accelerated their distribution (Bösch 2011, 128–142). The piecemeal use of photography led to fundamentally new conjunctures of text and image and facilitated considerably expanded potentials for layout and new journalistic formats (Zimmermann and Schmeling 2006).

While countless representations of journalistic practices at present and in history largely ignore technological aspects and factors, a perspective on the 'impacts' of technologies is virtually impossible. It is thus necessary to take into account economic, social and cultural determinants of the history of usage of relevant technologies. Although the implementation of digital computer technology decisively changed the overall process of the production of media and especially journalistic practices over the last decades, the yet unanswered question is who used it, where it was used and for what motives. This volume hence approaches this complex from an interdisciplinary perspective and aims to contribute to its analyses of the manifold interplay between science and technology on the one hand and between society, culture and politics on the other.

These correlations imply that the causal relationship between technologies and their effects is distinguished, that both the production and sphere of consumption count in the history of applied technologies, and that not only continuities but also discontinuities should be observed in the implementation of technologies (Boczkowski and Lievrouw 2008). In this way, the volume ties in with the discussion of the social construction of technologies, which was directed against technological determinism. This discussion worked out that 'relevant social group' and 'interpretative flexibility' are among the key concepts of such an approach (Bijker 2008). In the German research context, the concept of 'appropriation' (Aneignung) was strengthened (though in relation to audience and not journalism research). Appropriation is to be understood as the complex historical process during the course of which new media and technologies are incorporated into standards of conduct, in which those perceptions and requirements of the users are adopted that are understood as acting subjects (Schmidt 1998).

1 The Current State of Research

In the history of media so far, the question of the technological implications of journalistic professional practice assumes in a positively striking way only a very limited significance. Neither for the 19th nor the 20th century have studies on the history of media paid even remotely sufficient attention to the question of technologies in the workplace and journalists' interaction with communications systems such as telegraph, telephone, teletypewriter and Internet. Research focuses on questions of political context, commercialization processes, cultural appropriation of American journalistic styles in Europe (investigative practices) and the political self-conceptions and dependencies of journalists (Zimmermann 2006; Hodenberg 2006). Media history has so far largely bypassed above all the massive economic and technological upheavals of the last decades, right up to today's medial applications (Agar 2005). However, general acceleration and globalization processes of medial communication were worked out and the nationally diversified parameters of journalistic practice repeatedly highlighted in the history of mass media (Barbier and Bertho Lavenir 2000; Requate 2010).

Within communications studies and journalism, on the other hand, which are open for the question of the actual impact of technological innovations on journalistic work and the appreciation of technologies by journalists, but where the question is also posed as to how important technologies as a whole are in the alteration of journalistic practice (Cottle and Ashton 1999), historical perspectives play practically no role at all. There are, however, striking exceptions: Especially the effects of the acceleration of global information flows and the alteration of the forms and content of newspapers by new means of communication have been addressed on different occasions. Such effects emerged by no means automatically, and the 'old' and the 'new' forms of reporting stood side by side. This means, at the same time, that earlier linear models of modernization in favor of more complex perspectives were abandoned (Bonea 2010). For the USA, the links between technological developments, the emergences of news agencies and, as a result, new forms of cooperation on the part of newspapers were highlighted (Blondheim 2000). It was likewise demonstrated for the USA how the industrialization and professionalization of newspaper journalism prepared current developments in the arena of news work and the convergence of media against the backdrop of the growing economic concentration of media enterprises. The history of technology is incorporated from this perspective into the history of growing institutional differentiation and division of labor in media (Nerone and Barnhurst 2003).

Current research shows that journalists strongly perceive changes in their work as being determined by technology, since they are confronted by it directly and on a daily basis in their workplace (Örnebring 2010, 58). It equally demonstrates a whole suite of studies that interpret changes predominantly as technology-driven (McNair 1998, Pavlik 2000, Welch 2000). These studies therefore follow a perspective of technological determination and stress economic efficiency. Applied to the area of media and journalism, this deterministic model that was disseminated in the USA supported a perspective according to which technology substantially determines the form in which content is presented.

Conversely, European and in particular German research has underestimated the technological basis of the media. This tendency is simultaneously supported by the traditional dichotomy of research into media and communication: Technological developments are subjects of engineering and the natural sciences, while dimensions of content are studied by the social sciences and humanities. This dichotomy, however, has been relativized with time. Now, in their definitions of the notion of 'media,' communication studies—traditionally more concerned with the person specification (work requirements) than with technological aspects of the journalistic profession—relate the media's characteristics to the state of the art of technologies used for the production and distribution of information and entertainment offers (Kleinsteuber 1992, 305; Weischenberg 1995, 15–16). Since the mid1980s and against the backdrop of an accelerating advancement of digital computer technology, media technology increasingly became the subject of media and communication studies. On the one hand, new technical systems and their effects on the editorial process were considered; on the other it was attempted to estimate future trends in journalism via surveys of users

(Weischenberg 1978, Prott 1983, Mast 1984, Weischenberg, Altmeppen and Löffelholz 1994, Weischenberg, Malik and Scholl 2006).

The results of such studies were quite contradictory. While some observed the disintegration and 'Taylorisation' of journalism, others stated that new technologies allowed for more autonomy and 'holistic' use (Haas 1999, 77–78). Many (meanwhile themselves historic) future scenarios turned out to be wrong. Yet it is worth noting that the findings of media and communication studies regarding the mechanization are often just snapshots and thus quickly relativized or superseded by new developments.

Research into media history that principally is in a position to put into perspective such deficits, however, is still dominated by a far-reaching blindness regarding technology. As elements of research, technical structures and actors are at least often not related in such studies. Mutual interrelationships are not sufficiently acknowledged or merely implicitly suggested. Ultimately, however, there are studies that explicitly consider media-technological developments—for instance in the area of printing and (media) agency technologies—from historical perspectives. These studies partly remain in the realm of listings and descriptions of media-technological innovations (for example Gerhardt 1986). Other studies embed the proliferation of new information and communication technologies more strongly in their respective cultural, economic and social contexts, yet at least in part appear to derive their findings from implicated medial properties rather than from empirical data (for example Giesecke 1991).

Of the newer publications in the area of journalism research that explicitly engage with the relationship between journalism and technology, Mark Deuze's study on working conditions in media companies (Deuze 2007) and in particular Patricia Dooley's overview of technologies in journalism in historical perspective (Dooley 2007) are especially noteworthy. Although the latter concentrates almost exclusively on the US and overstresses specific individual developments, it remains the first to acknowledge the specific social and cultural environment in a broader historical perspective. Equally good insights into the relationship between the 'new journalism' and the world wide web as well as between broadcast technology and journalism since the 1950s are offered by Martin Conboy's study on Britain (Conboy 2004).

A prospective new approach to the problem has been introduced by Henrik Örnebring (2010), who reverts to approaches originating in 1970s Labour Process Theory. Örnebring studied the interrelationship between journalism and technologies by focusing on changes in journalism as 'work' and 'profession' and the standards and qualifications associated with these changes. Since several newer studies within contemporary journalism research are concerned with how technologies are restructuring and transforming journalistic practice (e.g. Boczkowski 2004; Deuze 2007; Neuberger, Nuerbergk and Rischke 2009), this perspective equally has potential benefits for the analysis of recent and present developments in the area of journalism. In this context, the necessary historicization of the topic is striking. Even before the general digitalization, media technologies as well as interactive videotext constituted alternatives to print products, though they by no means gained such massive acceptance as their potential might have suggested (Boczkowski 2004, 19–50). As newer historical case studies demonstrated, massive tendencies of computerization already took place from the 1970s onwards and formed the basis for today's digitalization. With the replacement of film by electronic picture recording, the introduction of text-processing and content management systems as well as desktop publishing technologies, editors had to first of all accustom themselves—in view of the demands of multitasking—to working with the new graphical surfaces. These processes were evaluated by the journalists themselves as both an expansion of research possibilities and a growth in autonomy, at least in principle. It was not the technologization as such that was scarcely criticized by journalists but rather the rationalization imperative that was behind its introduction. The cutback in fixed positions and time pressures constitute far greater problems than the new technology (cf. Reddick and King 1997; Egnolff 2010; Gränitz 2010; Raubenheimer 2010).

2 Approaches and Aims of the Volume

The volume 'Technology, Media and Journalism' picks up such newer perspectives and is geared towards developing further research at an international level. It departs from the premise that new technologies are not effective sui generis, but on the contrary develop their specific potential only within the cultural and social contexts in which they are applied and through which they are mutually interrelated.

In particular, this book aims to introduce the great potential of historical knowledge into the research area of Technologies and Journalism. For the first time in this field, the volume brings to the fore consistently historical perspectives that go back to the 19th century. Corresponding with the basic assumption that history reaches into the present, contemporary developments will be considered as part of the historical perspective.

Based on these underlying assumptions, previous processes in the field of journalism and technologies will be related to more recent and present developments. In general, one can demonstrate that journalism itself has always been innovative and was able to meet challenges—regardless of whether these were induced either by technological developments or by the transformation of the economical, socio-cultural or profession-political context. The following general criteria meet these observations and try not to act on the assumption of a unidirectional technological determinism, but to represent the complex and multifactorial set of conditions with respect to professional journalism over the last one and a half centuries.

(1) *Separation and immediacy:* Since the development of professional journalism in the 19th century, the technological basics of media have always influenced the work of journalists. In this context, the separation between printing technology and the work of news gathering and editing was in the first instance fundamental for the emergence of the journalistic profession. Up to the present time, editing technology gained more and more immediate influence on the work of journalists because the number of mediating entities was reduced. Important steps in this process were the immensely increasing news flow due to telegraphy, the emergence of local news reporting around the turn of the 20th century, and the introduction of the teletype machine/ticker in the 1930s.

(2) *Speeding up the news process:* Since the mid-19th century, newsmaking has been sped up tremendously. The discourse of speed—understood as 'at heart a capitalist logic of competition and use of technology to increase productivity' (Örnebring 2010, 65)—has become a wholly naturalized element of journalism and forms a template for how journalists understand new technologies, i.e. primarily as an opportunity to speed up the news process (Hampton 2004). There is a wide spectrum of significant developments in this area, ranging from the speeding up of post and passenger traffic, telegraphy, telephony and teletyping to digital communication media of recent times. Notably, the increasing speed required news to be concise and brief, related to their extent as well as their linguistic complexity.

(3) *Mobility:* Technologies whose primary function was to transcend space made journalists' work less mobile and more individualistic. Thanks to the telegraph, and later the telephone and the teletype machine, journalists did not need to leave their desks to get the news. Along with that, their desk with typewriter and, later, the computer as a workplace gained increasingly in importance. Based on Boyer's (2004) study of the role of the typewriter in the mechanization of office work, primarily by spatially linking the office worker to the journalist's desk, one can speculate that the typewriter might have been used for similar purposes in newsrooms. Indeed, MacGregor (1997) attributes exactly this function to the computer, turning journalists into 'mouse monkeys,' bound to their computers and tasked with repackaging incoming information into a variety of multimedia content.

(4) *'Taylorization' vs. 'Autonomization':* Although gaining more control over the process of media production has always been an important intention of the implementation of new technologies in the editorial departments, there is no clear tendency towards a 'Taylorization' or 'Autonomization' of journalists' work. As several studies (Pavlik 2000, MacGregor 1997, Deuze and Paulussen 2002) have shown, the introduction of new computer hard- and software required journalists to become more skilled in carrying out technical tasks and this led to new dependencies in the production of media that may undermine the autonomy of journalists and give more power to editors. However, the new technologies remain always an instrument/a tool that has been adapted to the specific socio-cultural contexts of media. Do we perhaps think of the antagonism 'Taylorization' vs. 'Autonomization' only because the work of journalists—since the origins of professionalization—always had to submit to certain constraints (as for instance the political orientation of newspapers and media enterprises)?

(5) *News gathering vs. news processing/editing:* Nevertheless, there is a tendency towards an increasing importance of news processing (including the visual composition) in place of news gathering and investigative reporting. In some cases, this may result in a polarization of journalism as a profession: A few journalistic 'stars' (most frequently TV news presenters, but some newspaper columnists have also acquired 'star' status) who command public recognition carry out the investigative research on-site, whereas the majority of journalists performs routine tasks in the newsroom (Ursell 2004). At the same time, editorial departments use more and more pre-processed informa-

tion, such as news agency material, and simply process it. It is remarkable that on-site investigations were not possible until technological innovations like railway connections, telegraphy and telephony had been established in the early times of professional journalism (Bösch 2011, 109–112).

(6) *Upskilling, Reskilling and Deskilling:* In media history, it has been worked out in manifold ways how the profession of the modern journalist emerged in the 19th century. This professionalization process was based on the establishment of special qualifications as well as on the organization in professional associations and on media achieving a certain autonomy vis-à-vis the politics of the day, although journalists continued to be understood as representatives of political positions. Accompanied by the establishment of professional standards and ethics, by the differentiation of newspaper categories and the emergence of research, the career journalists obtained specific skills and an independent professional profile as a result of the ever greater significance of news collection (Requate 1995). This was the consequence of training on the job, though there were scarcely any formal training courses. Today, and since the 1970s, a paradigmatic shift can again be observed that appears to be just as significant as that of the 19th century. It shows very clearly that in the field of journalism, processes of upskilling, reskilling and deskilling are occurring simultaneously. Journalists work with all the segments of the news-gathering process from investigation to production (and thus have a broader skill base), but production takes more time and journalists find less need for skills associated with actively seeking out and investigating news (Örnebring 2010, 67; Nygren 2008, 164–166). As journalists become more skilled in digital production techniques, they may find less use for their news-gathering and collating skills. In this regard, recent developments are particularly impressive, but in a historical perspective, processes of upskilling, reskilling and deskilling could always occur whenever new technologies were introduced in media companies (telephone, typewriter, teletype machine, photography). In general, the necessary skills for journalistic work are subject to continuing changes, modifications and transformations due to not only the development of media technologies but also their application in different cultural and social contexts.

Blogs are a symbol of today's transition from unidirectional mass media, which still constituted the reigning paradigm in the 1960s, to a participative form of mass media. Just how strong the different forms of blogs (diary style blogging, filterblogging, topic-driven blogging) influence today's journalistic practice is a current research question. Bloggers function at least in exceptional cases as 'independent journalists,' they supply a form of 'citizen journalism' in the sense of gatewatching and 'counter-public sphere.' Alone as a result of their fragmentary and personal-reflexive literary form and due to their limited reach, however, they do not replace the classical standards and achievements of quality journalism (Rettberg 2009, Engesser and Wimmer 2009).

(7) *Continuities and breaks in professional role models, self-images and the epistemic systems of journalistic practices:* Although journalistic labor—and the necessary skills of journalists—has changed massively in the face of technological evolution, key elements of the professional role model and the self-images of journalists have remained quite stable. These include journalistic independence and autonomy; verifiability, accuracy and consistency of reporting; importance and social relevance of news. These elements of a professional self-image go back to the 19th century (Requate 1995) and stabilized—with several turns and transformations (Hodenberg 2006)—over the past 150 years. Thus, this dimension of a professional history is not essentially related to technology and has been characterized by a high degree of continuity, at least until recently.

In this context, reference should be made to the internationally fierce debate on quality in journalism. Current professional quality standards and the social status of professional journalism appear to be called into question by the advancing digitalization and the emergence of blog forums. It was hitherto undisputed that the differentiation of departments, a network of correspondents, the explicit addressing of the viewers and readers, as well as a broad content of coverage were all aspects of quality journalism. Now, in the age of growing expansion of online media, must one now let only the public decide on what constitutes quality? Have the new communications possibilities of the Net led to journalism losing in standing? Has the growing pressure of being up to date in important areas of journalism led to a reduction in the level of quality and is, therefore, 'the century of journalism over,' as the researcher Siegfried Weischenberg—perhaps too skeptically—claimed? (Weischenberg 2010, cf. Lünenborg 2012). Or is journalism—confronted by the Internet—on its way from its traditional role of gatekeeping to a participatory journalism, but remains in its societal role? Users and readers are quantitatively and qualitatively overwhelmed with perceiving journalistic tasks themselves (Neuberger 2000, Neuberger and Quandt 2010).

(8) *Visuality:* From the mid-19th century onwards, a rapid change of visual systems took place in the field of media. Outstanding innovations in this regard include the possibility to print pictures by halftone printing methods, the emergence of photo agencies, improvements of printing techniques as well as—in recent times—digital image processing and storage (including digital photo archives). In addition to the tendency towards a growing number of photos and other visual elements in print media, the introduction of television as an audio-visual medium was another key step in this development. Since the 1990s the multi-medial Internet, which is based on digital computer technology, has joined these media. The aforementioned trend towards brevity—in this case by the seemingly quicker perception and the simplification of the consumption of visual information—and towards the use of pre-processed information (news agency material) can also be observed in this context.

This volume offers the opportunity to approach these eight criteria from new, including historical, perspectives and to reach more balanced conclusions. In order to do so it is necessary to consistently evaluate those transformation processes that accompany technological changes in terms of their actual influence on everyday journalistic practices. This equally offers an opportunity to deconstruct as well as to position socially and historically the findings by contrasting them with journalists' self-perception of their roles.

The emergence of such innovative perspectives and approaches is a specific concern of this book. Unquestioningly, the past 15 years have produced cataclysmic changes in the area of journalistic practice. In the context of this volume, such relevant developments and processes will not be considered in isolation or from within their disciplinary premises, but within an overall context that combines and integrates historical, technological and media aspects. This allows for the comparison of contemporary and historical media developments, the emergence of specifics and commonalities as well as the identification of continuities and ruptures or phases of rapid innovation.

In order to achieve this aim, this volume brings together participants and international expert delegates from the areas of media- and communication studies on journalism, technology and media sociology, and technology and media history. This provides an international research context that combines the different disciplines. In the face of deterministic technology perspectives on the one hand and a missing acknowledgement of the technological dimension on the other, it is imperative to provide journalism research with a new dynamic through new and integral approaches. With direct references to the everyday work life of journalists, the results may also be of great relevance for research on contemporary journalistic practices.

3 The Chapters

The following chapters consider not only the implementation and achievements as well as failures of media-technological innovations themselves, but in particular the influences on the practical demands (combined with the question of the necessary skills) and processes in media companies as well as professional roles and social positions/self-perception of journalists.

(1) The first part of the book contains four articles that unfold the overall problem of the interactivity of journalism and technology in light of historical developments and on the basis of theoretical reflection.

In 'Neither good, nor bad; nor neutral: The Historical *Dispositif* of Communication Technologies,' *Andreas Fickers* introduces fundamental concepts, not least the actant concept from Bruno Latour. If it is always a question of 'newness' in the history of technology, goes the author's argument, then this is to be disconnected from all ideas of progress—but also from threat scenarios. In the area of the interrelationship of technologies and journalism, implicit normativity is to be reflected, for example when innovations in the area of journalism are always defined as 'good' or 'bad.' The author presents one of the most decisive topics of the volume, namely the question of the acceleration of news flows, which are also embedded culturally and in an innovation history of infrastructures, so that it can be concluded that one cannot automatically talk of the 'effects' of new information technologies. Fickers points to the considerable significance of the discourse history of technological innovation when he recalls the futuristic positions and he emphasizes how the history of radio is to be understood both as an extension of communications spaces and as a domestication of the—by all means active—listeners. The history of communications apparatus, suggests the author, is also to be understood as the history of human perception in the sense of the *dispositif* theory (Jean-Louis Baudry).

In 'Temporalities of News: News and Speed from the Early Modern Era to the Present,' *John Nerone* likewise suggests a broad historical arc and also argues generically. Alongside 'speed,' he stresses 'newness' and 'news' as central analytical concepts and subjects of the history of journalism. He reflects fundamental connections between acceleration processes in the history of news and the way in which journalistic texts are introduced into the cultural spheres of every epoch. In a strict model of phases from the colonial era to the present, Nerone traces the forms and contents of newspaper journalism, its interpretative power and its practical uses for the reader. The history of the introduction of telegraphic news constitutes for him not only a process of acceleration but also a process of economic monopolization. The telegraphic news form, equipped with distinct characteristics, had irreversible consequences for the forms of political communication, and yet in a historical account the news environment always counts as well. The significant innovation of investigative journalism, according to the author, is not only to be regarded as typical as a social innovation but also as textual practice, in which a specific temporal structure is established. As a result, further temporalities altered the distribution of the telephone, the radio and the Internet but the journalistic task of selecting news and presenting it as essential and as authoritative remained.

In 'From Hand-made to Technology-driven Journalism: Changes and Developments in Research, Writing, Editing, and Make-up,' *Jürgen Wilke* advocates the thesis in a historical longitudinal analysis that the history of journalism can be written as the history of applied technologies. At the beginning, there are manual skills, practical and intellectual approaches with the fixed contents that need to be constantly re-arranged. From the simple printing mechanisms via the telegraph and the telephone to complete computerization, Wilke sketches a picture of advancing technologization and an acceleration of labor demands but which offer the journalists opportunities for individualization and growing competences. For systematic comparisons in the research field of journalism-technology, the approach of the author—following on from Dooley—of differentiating clearly between the different analytical levels of journalistic work can be particularly fruitful: research of information/news, processing and treatment of information/news, content/presentation of information/news, and the distribution of information/news. From these levels, the individual technologies and their impact can be observed in a nuanced way. It is also visible how sensible it is to assume an open and broad concept of technology.

In 'The Interpretive Turn in News,' *Kevin G. Barnhurst* goes back to the early 20th century, at a time when explanatory journalism emerged with the new American type of news magazine. The article places overall innovations in the history of journalism, the history of technologically-based individual media and political self-conceptions and conditions in a close relationship to one another. Barnhurst regards the beginning of interpretative news as a reaction to advancing technologization of news journalism and also the continuing medialization of society as a whole. Precisely from the wealth of the hitherto treasured facts, the idea arose of the social responsibility of journalists, who began to assume a political role. The powerful medium of television, on the other hand, tended to only show what was concrete and visible, and placed the audience in a passive role by means of its overwhelming visuality. Ultimately, however, Barnhurst also observes an interpretative turn here. Finally, the author examines the fundamental modus of journalism, its 'literary' characteristics and its 'realism,' and sounds out the chances of an interpretative-orientated journalism in the age of the Internet.

(2) The use of image and visualization technologies in the media and journalistic context is the central concern of the second section. The articles in this section treat several aspects of visualization with a focus on photography.

In 'Contemporary or Commercial? The Impact of Innovative Image Reproduction Techniques on the Imagery of Nineteenth Century Mass Media in Germany,' *Rita Gudermann* addresses images as goods on capitalist and international markets since the 19th century. The ground was already laid in the late 19th century for today's broad availability of digital stock photos. Innovations of reproductive techniques played a decisive role here. On the one hand, they made it possible to print illustrations more quickly, in an ever larger print run and also, increasingly, in color. On the other hand, this resulted in higher costs and these had to be made good not only by means of higher print runs but also through multiple use. Though magazine editors claimed to offer a certain up-to-dateness, they chose such stock pictures for their periodicals (and books) that met the needs of 'stockability.' Using the example of the international picture trade and, especially, two publishing houses in Leipzig, the fundamental options of the publishers between a syndication strategy and up-to-dateness are shown. Thus, the potential of the new illustration techniques was realized in the context of growing economic internationalization and by means of the publishers accommodating the massive interest of the public for illustrations. In today's accounts and interpretations of the flood of digital images, this historical dimension is generally overlooked.

In his article on 'Eyewitnesses? The Visual Depiction of Events around 1900,' *Jens Jäger* also turns his attention to the distinct interest of consumers in abundant and, if possible, current illustrations. He goes back to the 19th century with its illustrated family magazine *Die Gartenlaube* (The Garden Arbour) and connects his pictorial historical approach with the question of the growing importance of media events. Via the telegraph and the news agencies, the press was in a position to report extraordinarily swiftly on current affairs, which were staged as 'events,' but it proved to be difficult to also print current images of these events. For one thing, such material was lacking in many cases or it was delivered too late. For another thing, the dramatic visualization was insufficient. Thus, it is hardly surprising that the visualization of a great media event such as the sinking of the *Titanic* happened in various ways: Drawings played an important role and photographs were not necessarily more important; furthermore, use was made of available material such as maps and diagrams. 'Events' could emerge for the first time via the media and their dramatic visualization, as the author argues, and illustrations guaranteed at the same time 'authenticity.' Visual event journalism and the primacy of news allowed the reader to become an 'eyewitness.' It seems as though the involvement of television viewers was already laid out in these forms of presentation on the part of press photography.

(3) In the third section of the book, current problems of the interrelationships of technological innovations and journalistic practice are addressed with recourse to standards that have evolved historically and to reflexivity. It is a question of the work of journalists, the relations between technology and the profession of journalism as well as the professional roles of journalists. Central to this is the pressing (and also historically inherent) problem of 'quality'—without quality control, there cannot be a professional journalism and will not be one in the future. The question will be pursued as to whether and how the professional journalists can secure the profile and quality of their work under conditions of considerably accelerated, globalized flows of news and in view of an audience that is active even by the standards of news procurement and interpretation. In this way, standards of journalism are debated that emerged in the context of a historical setting of technological practices and the public sphere, which has since changed considerably. These developments especially throw up the question of how journalists conceive their own professional role today and what the case is with the oft-celebrated 'participatory' public communication. Therefore, the volume also picks up

on current trends and not only critically illuminates them on the secure basis of qualitative and quantitative research approaches but also places them in an international context.

It cannot be overlooked, as *John Pavlik* explains in 'Mobile, Networked and Digital Technology. Implications for Journalistic Work,' that these technologies fundamentally influence the work of journalists today. They enabled the journalists, as the author continues with reference to Ithiel de Sola Pool, to collect news efficiently and cost-effectively and to distribute it. However, so-called citizen journalists also use mobile communications appliances in order to circulate video images and other content via their own networks as well as traditional news channels. The downsides of this 'paradigmatic shift'—in the terminology of Thomas S. Kuhn—cannot be overlooked, as Pavlik argues very clearly. They consist of news being circulated by the activities of 'networked individuals' (William Dutton) and by professional journalists in many cases not having access to this news. Therefore, it cannot be examined for verisimilitude and context. One is dealing here with an enormous acceleration tendency, but also with social and political interests that stand behind the news content and can barely be overlooked. Furthermore, it is to be kept in mind, according to the author, that editorial staffs have been reduced for a long time now and that, through the convergence of computer science and computational journalism, 'algorithmic' approaches in professional journalism massively expand, which allow the quasi automatized production of texts. In the current situation, the fundamental question emerges as to how storytelling, the essential form of journalism, and the tools of a journalistic work that aims at credibility are to look in the future.

Tobias Eberwein seamlessly picks up this question of journalistic quality and uses for this purpose the instrument of a comparatively applied survey research. Central to his article on 'Journalistic Quality as Crowd Wisdom? What Journalists Think about Criticism on the Social Web' is, for one thing, the question of media self-regulation as a form of non-state media accountability. This means that the author asks how journalists view themselves and reflect their own professional practice. For another thing, the author introduces the problem of how the activated audiences judge and comment on journalistic work. Are such 'participatory' forms really suited to balancing out the recognized deficits of medial self-regulation and journalistic quality management? How do journalists perceive, on the other hand, their new cyber watchdogs? Essentially, Eberwein observes quality in today's blog journalism (more so in the USA than in Germany) and to this the fact belongs that the users of such blogs could be motivated to rethink the quality criteria of journalism. Thus, the national differences of journalistic cultures are to be incorporated into the setting and the academic interpretation. In this way, a differentiation takes place between 'liberal,' 'democratic-corporatist' and 'polarized-pluralist' systems and cultures, to which are added the context of transformation countries such as Estonia, Poland and Romania, as well as Jordan and Tunisia, which are only just at the beginning of a stronger free and self-regulating media system. It becomes clear in many places that the criteria of journalistic quality are strongly dependent on historically inherent political and cultural contexts. This means that these contexts are capable of differentiating unifying technological action conditions and trends.

In 'Quality in Journalism Reconsidered: The Limits of Realism,' *Mitchell Stephens* approaches the problem of quality journalism with a strong focus on historical reflections. It can currently be observed, according to the author, that classical standards are meanwhile very quickly threatened by job cuts. For another thing, through the Web, the activities of amateurs and algorithmic production systems the amount of newsworthy events is multiplied. As a result, the almost overwhelming problem arises of how journalists are to select the news under these conditions and interpret it for their audience. The maxim that journalists should be 'going places' does not bring us any further. How, therefore, do they achieve distinct quality in view of a public that not only expects speed but, in many cases, is also no longer prepared to pay money for the news? At this point, the author resorts to historical analogies and discourses. The 'realistic' painting of the 19th century was already focused on 'truth' and richness of detail, but the photography fundamentally challenged such painting and devalued it. Therefore, as Stephens concludes, the task of today's journalism is no longer so much to collect and examine 'facts' and 'details,' as they formed the basis of the simplified concept of objectivity of applied social and natural sciences in the 19th century. Instead, the task is located in the 'depth,' just as the impressionists had already searched for it. Journalistic work should consciously distance itself from an ostensible concept of realism and proceed in the direction of 'wisdom journalism.'

Thus, some arguments can be discerned here that were already presented by Kevin G. Barnhurst. *Christoph Neuberger* addresses the fundamental statement in the same direction; methodically-speaking, his approach is related to that of Tobias Eberwein. In his article 'User Participation and Professional Journalism on the Internet: Theoretical Background and Empirical Evidence,' Neuberger stresses more strongly than the other authors the importance of existing media ensembles and as a result coexisting or 'complementary' professional practices in journalism. The author also sees no fundamental difference between traditional and blog journalism, especially since historically-speaking forms of reader participation already existed—and in the 'new virtual world,' he continues, levelled hierarchies exist. Empirical research now shows that today's audiences are very aware of the question of quality. The websites of the official media, but also Wikipedia, receive better evaluations for informational quality than other pages. The official media are likewise evaluated highly when it comes to an interest in obtaining an overview and orientation. Whoever searches for entertainment in the media prefers social media, but the professional journalists also drew massively on online services and web blogs in order to find out about topics and trends. In this relatively harmonizing view on the basis of material from Germany and the USA, whose applicability to other media cultures would still have to be examined, the problem then also resulted regarding how professional journalism can in the future satisfy its 'customers' through quality.

Finally, the article by *Karin Wahl-Jorgensen* on 'Changing Technologies, Changing Journalistic Epistemologies: Public Participation, Emotionality and the Challenge to Objectivity' focuses on the question of the interrelationship of media technologies and participatory medial practices. At the same time, the article contains fundamental considerations on the construction of the relationship of public and media users, which confer on the piece not only an exploratory but also a summing-up character. The author reaches by means of her postulate of an 'affective turn' in journalism research a completely different paradigm. It appears suited to uncovering the background features of 'participatory journalism' that were so far unknown in this way—and it implies ethnographical research methods that could be followed analogously in a cultural history perspective and expanded on. According to this, it is clear that the 'rational' reader subject presumed in the history of European ideas and in Habermas' concept evidently never corresponded to reality. With modernity, popular culture also became societally and politically relevant, and through it emotions became, so to speak, legitimate. As Mitchell Stephens already noted, classical journalism was governed by the belief that 'facts' and 'opinions' could be separated from one another. This question positively calls for epistemic systems of journalism to be examined, and for them to be understood as rules, routines und institutionalized procedures that, in turn, are constantly influenced by historical power relations and professionalization processes. Between changing technologies and participatory forms (classical: letters to the editor and calls from radio listeners), new constitutive interrelationships continually emerged. Thus, today's appreciation of 'authenticity' and nearness to the audience builds on older practices and continual discourses. The account of the author aims lastly at the character of 'subjective journalism' and the new forms of journalistic portrayal, which are implied by means of the postulate of the inclusion of emotions in the research field.

Taken together, the temporal spectrum of the book ranges from the use of technical communication media, such as telegraph and telephone as well as the professionalization of print technologies in the 19th century, to the contemporary phase of completely digitized communication and production systems. A central focus is on text and image journalism in the area of print media, but radio and television and then the challenge of the Internet are also considered, as well as the point of media ensembles. The complete analysis includes a timeframe of more than 150 years and thus a continuum of technological, social, cultural and media developments in the area of journalistic history that in this form has so far not been studied. Overall, the volume demonstrates how technologies have influenced journalistic practices historically and currently in a multitude of ways, and how specific communications technologies underlie journalistic work processes and the media as a whole. Its quality as an actor itself, however, depended on economic and cultural historical conditions (and the same applies principally for the present), whose consideration must contradict all technological determinism. The authors of this volume follow less very concrete 'empirical' topics and instead target the justification of fundamental categories in the professional field of journalism and in journalism research. To this extent, this volume constitutes an empirical, historical and theoretical-reflexive project.

Works Cited

Agar, Jon (2005). Medium Meets Message: Can Media History and History of Technology Communicate? *Journal of Contemporary History*, 40 (4), 793–803.

Barbier, Fréderic, and Catherine Bertho Lavenir (2000). *Histoire des médias de Diderot à internet*, Paris: Armand Colin.

Bijker, Wiebe E. (2008). Social Construction of Technology. In Wolfgang Donsbach (ed.). *The International Encyclopedia of Communication*, XI, 5031–5036. Malden, Oxford, Carlton: John Wiley & Sons.

Blondheim, Menahem (2000). The Click: Telegraphic Technology, Journalism, and the Transformations of the New York Associated Press. *American Journalism*, 17 (4), 27–52.

Boczkowski, Pablo J. (2004). *Digitizing the News: Innovation in Online Newspapers*. Cambridge, Mass.: MIT Press.

Boczkowski, Pablo J., and Leah A. Lievrouw (2008). Bridging STS and Communication Studies: Scholarship on Media and Information Technologies. In Edward J. Hackett et al. (eds.). *The Handbook of Science and Technological Studies*, 949–977. 3rd ed. Cambridge, London: MIT Press.

Bösch, Frank (2011). *Mediengeschichte. Vom asiatischen Buchdruck zum Fernsehen*. Frankfurt am Main, New York: Campus.

Bonea, Amelia (2010). The Medium and its Message: Reporting the Austro-Prussian War in the Times of India. *Historical Social Research*, 35 (1), 167–187.

Boyer, Laura K. (2004). "Miss Remington" Goes to Work: Gender, Space and Technology at the Dawn of the Information Age. *Professional Geographer*, 56 (2), 201–212.

Conboy, Martin (2004). *Journalism: A Critical History*. London, Thousand Oaks, New Delhi: Sage.

Cottle, Simon, and Mark Ashton (1999). From BBC Newsroom to BBC Newscentre: On Changing Technology and Journalist Practices. *Convergence: The International Journal of Research into New Media Technologies*, 5, 22–43.

Deuze, Mark (2007). *Media Work*. Cambridge: John Wiley & Sons.

Deuze, Mark, and Steve Paulussen (2002). Online Journalism in the Low Countries: Basic, Occupational Characteristics of Online Journalists in Flanders and the Netherlands. *European Journal of Communication*, 17 (2), 237–245.

Dooley, Patricia L. (2007). *The Technology of Journalism: Cultural Agents, Cultural Icons*. Evanston, Ill.: Northwestern University Press.

Engesser, Sven, and Jeffrey Wimmer (2009). Gegenöffentlichkeit(en) und partizipativer Journalismus im Internet. *Publizistik*, 54, 43–63.

Egnolff, Mareike (2010). Die Veränderungen der Berufsbilder in der Printpresse. Die Saarbrücker Zeitung. In Clemens Zimmermann (ed.). *Medienpolitik und mediale Strukturen, 1955–2005 (Medienlandschaft Saar von 1945 bis zur Gegenwart, Vol. 2)*, 363–389. München: Oldenbourg.

Gerhardt, Claus W. (1986). *Drucken in Geschichte und Gegenwart*. Frankfurt am Main: Polygraph.

Giesecke, Michael (1991). *Der Buchdruck in der frühen Neuzeit. Eine historische Fallstudie über die Durchsetzung neuer Informations- und Kommunikationstechnologien*. Frankfurt am Main: Suhrkamp

Gränitz, Barbara (2010). Die Veränderungen der Berufsbilder im Fernsehjournalismus. Der saarländische Rundfunk. In Clemens Zimmermann (ed.). *Medienpolitik und mediale Strukturen, 1955–2005 (Medienlandschaft Saar von 1945 bis zur Gegenwart, Vol. 2)*, 321–361. München: Oldenbourg.

Haas, Hannes (1999). *Empirischer Journalismus. Verfahren zur Erkundung gesellschaftlicher Wirklichkeit.* Wien, Köln, Weimar: Böhlau.

Hodenberg, Christina von (2006). *Konsens und Krise. Eine Geschichte der westdeutschen Medienöffentlichkeit 1945–1973.* Göttingen: Wallenstein.

Kleinsteuber, Hans (1992). Zeit und Raum in der Kommunikationstechnik: Harold A. Innis' Theorie des technologischen Realismus. In Walter Hömberg and Michael Schmolke (eds.). *Zeit, Raum, Kommunikation*, 319–336. München: UVK.

Lünenborg, Margreth (2012). Qualität in der Krise? *Aus Politik und Zeitgeschichte*, 62 (29–31), 3–8.

MacGregor, Brent (1997). *Live, Direct and Biased? Making Television News in the Satellite Age.* Oxford: Oxford University Press.

Mast, Claudia (1984). *Der Redakteur am Bildschirm. Auswirkungen moderner Technologien auf Arbeit und Berufsbild des Journalisten.* Konstanz: Universitätsverlag.

McNair, Brian (1998). *The Sociology of Journalism.* London: Bloomsbury Publishing.

Nerone, John, and Kevin G. Barnhurst (2003). US Newspaper Types, the Newsroom, and the Division of Labor, 1750–2000. *Journalism Studies*, 4 (4), 435–449.

Neuberger, Christoph; Christian Nuernbergk, and Melanie Rischke (eds.) (2009). *Journalismus im Internet: Profession—Partizipation—Technisierung.* Wiesbaden: VS Verlag für Sozialwissenschaften.

Neuberger, Christoph (2000). Renaissance oder Niedergang des Journalismus? Ein Forschungsüberblick zum Online-Journalismus. In Klaus-Dieter Altmeppen, Hans-Jürgen Bucher, and Martin Löffelholz (eds.). *Online-Journalismus. Perspektiven für Wissenschaft und Praxis*, 15–49. Opladen: Westdeutscher Verlag.

Neuberger, Christoph, and Thorsten Quandt (2010). Internet-Journalismus: Vom traditionellen Gatekeeping zum partizipativen Journalismus? In Wolfgang Schweiger and Klaus Beck (eds.). *Handbuch Online-Kommunikation*, 59–79. Wiesbaden: VS Verlag für Sozialwissenschaften.

Örnebring, Henrik (2010). Technology and Journalism-as-labour: Historical Perspectives. *Journalism*, 11 (1), 57–74.

Pavlik, John (2000). The Impact of Technology on Journalism. *Journalism Studies*, 1 (2), 229–237.

Prott, Jürgen (1983). *Berufsbild der Journalisten im Wandel. Zeitungsredakteure unter den Bedingungen der Bildschirmarbeit.* Frankfurt am Main: Rita G. Fischer.

Raubenheimer, Tina (2010). Die Entwicklung der Berufsbilder im Hörfunk. Der Saarländische Rundfunk. In Clemens Zimmermann (ed.). *Medienpolitik und mediale Strukturen, 1955–2005 (Medienlandschaft Saar von 1945 bis zur Gegenwart, Vol. 2)*, 277–320. München: Oldenbourg.

Reddick, Randy, and Elliot King (1997). *The Online Journalist. Using the Internet and Other Electronic Resources.* Fort Worth: Thomson Learning.

Requate, Jörg (1995). *Journalismus als Beruf. Entstehung und Entwicklung des Journalistenberufs im 19. Jahrhundert. Deutschland im internationalen Vergleich.* Göttingen: Vandenhoeck & Ruprecht.

Requate, Jörg (2010). Kommunikationsmedien und Gesellschaft. In Reinhard Sieder and Ernst Langthaler (eds.). *Globalgeschichte* 1800–2010, 439–469. Wien, Köln, Weimar: Böhlau.

Schmidt, Uta C. (1998). Radioaneignung. In Inge Marßolek and Adelheid von Saldern (eds.). *Zuhören und Gerhörtwerden: I. Radio im Nationalsozialismus. Zwischen Lenkung und Ablenkung*, 243–360. Tübingen: Edition diskord.

Ursell, Gill (2004). Changing Times, Changing Identities: A Case Study of British Journalists. In Torben Elgaard Jensen and Ann Westenholz (eds.). *Identity in the Age of the New Economy*, 34–54. Cheltenham: Edgar Elgar.

Walker Rettberg, Jill (2009). *Blogging (Digital Media and Society Series)*. Cambridge, Malden, Mass.: Polity Press.

Weischenberg, Siegfried (1978). *Die elektronische Redaktion. Publizistische Folgen der Neuen Technik*. München, New York: Saur.

Weischenberg, Siegfried (1995). *Journalistik. Theorie und Praxis aktueller Medienkommunikation. Band 2: Medientechnik, Medienfunktionen, Medienakteure*. Opladen: Westdeutscher Verlag.

Weischenberg, Siegfried (2010). Das Jahrhundert des Journalismus ist vorbei. Rekonstruktionen und Prognosen zur Formation gesellschaftlicher Selbstbeobachtung. In Gabriele Bartelt-Kircher et al. (eds.). *Krise der Printmedien: Eine Krise des Journalismus?*, 33–60. Berlin: De Gruyter.

Weischenberg, Siegfried, Klaus-Dieter Altmeppen, and Martin Löffelholz (1994). *Die Zukunft des Journalismus. Technologische, ökonomische und redaktionelle Trends*. Opladen: Westdeutscher Verlag.

Weischenberg, Siegfried, Maja Malik, and Armin Scholl (2006). *Die Souffleure der Mediengesellschaft. Report über die Journalisten in Deutschland*. Konstanz: UVK.

Welch, David (2000). News into the Next Century: Introduction. *Historical Journal of Film, Radio and Television*, 20 (1), 5–13.

Zimmermann, Clemens (ed.) (2006). *Politischer Journalismus, Öffentlichkeiten und Medien im 19. und 20. Jahrhundert*. Ostfildern: Thorbecke.

Zimmermann, Clemens, and Manfred Schmeling (eds.) (2006). *Die Zeitschrift—Medium der Moderne. Deutschland und Frankreich im Vergleich*. Bielefeld: transcript.

“Neither good, nor bad; nor neutral”: The Historical *Dispositif* of Communication Technologies

Andreas Fickers

Introduction

In 1985, Melvin Kranzberg, then professor for history of technology at the Georgia Institute of Technology and founding editor of the journal *Technology and Culture*, delivered the presidential address at the annual meeting of the Society for the History of Technology in which he explained what had already come to be known as Kranzberg's Laws—“a series of truisms,” according to Kranzberg, “deriving from a longtime immersion in the study of the development of technology and its interactions with sociocultural change” (Kranzberg 1986, 544).

The aim of this article is to reflect on Kranzberg's first law, which basically says that technology is—in sociological terminology—an actor category (Latour 2007).[1] Conceptualizing technology as actor category means to think about the agency of technology and technological artefacts and to reflect upon the social and cultural meaning of technology and what it does to those using it for specific purposes—for example for journalists producing television news, documentaries or radio features. As historian of technology and media, I'm especially interested in how technology is co-producing our perception of reality and how we can study and conceptualize the complex relationship between media technologies and mediated realities.

Rather than focusing in a more narrow sense on the role of technology in the field of journalism, this article reflects on the symbolic meanings inscribed into and attached to media technologies in order to demonstrate how the discourse of “newness”—that is the technological imaginary that accom-

1 The six laws are: 1. Technology is neither good nor bad; nor is it neutral; 2. Invention is the mother of necessity; 3. Technology comes in packages, big and small; 4. Although technology might be a prime element in many public issues, non-technical factors take precedence in technology-policy decisions; 5. All history is relevant, but the history of technology is the most relevant; 6. Technology is a very human activity—and so is the history of technology.

panies emerging technologies—affects the development and use of specific media technologies in journalistic practices, such as the transmission and use of telegraphic information for newspaper journalism or the making of a television documentary (Sturken and Thomas 2004).

As Patricia Dooley has convincingly argued in her book *The Technology of Journalism. Cultural Agents, Cultural Icons* (2007), the scholarship on the history of journalism reveals a bias that favours technology as driving force behind changes in journalism in general and news production in specific. Referring to Kranzberg's Law, one could argue that most works in the field of journalism studies are emphasizing the "good side" of technology, mainly by describing how technological devices and infrastructures such as the telegraph network and the news ticker, the mobile microphone and magnetic tape recorder, or the Sony Portapak and video editing have speeded up the production of news and helped to create a feeling of liveness and immediacy. Only a much smaller amount of works refer to the "bad" or more problematic sides of technology, for example by alluding to the constraints that processes of acceleration in the production, dissemination and reception of news embed in terms of quality of information and peoples' ability to reflect the ever growing mass of information they are confronted with.

But classifying past scholarship on technology and journalism by applying the dichotomy of "good" or "bad" would of course not do justice to authors such as Menahem Blondheim (1994), Brian Winston (1998), Paul Starr (2004), James W. Carey (2009), Richard R. John (2010), Michael Schudson (2011), or recently Thomas Birkner (2012), who all have rejected deterministic interpretations of the impact of new technologies on the practices of journalism. The emergence of the telegraph for instance provides us with a perfect example of what historians and sociologists of technology have coined as process of socio-technical "co-construction" (Hughes 2004; Pinch & Oudshoorn 2005). In their view, the development and use of new technologies is always the result of a complex process of negotiation in which a variety of social actors and institutions are involved. But the "circuit of technology," to use a term introduced by the cultural studies scholar Cynthia Cockburn (1992), does of course not end with the invention or development of a new technology. As Roger Silverstone (1994) has shown on the example of television, the process of domestication of a technology is as complex and contingent as is the process of invention. The "production-consumption junction," as Ruth Schwartz Cohen (1987) has convincingly argued, is the critical zone of convergence, in which a technology turns into medium or the use of a technological device becomes a social or cultural practice.

Based on a few concrete historical examples of emerging technologies this article wants to discuss a number of concepts and theories from the fields of media history, history of technology and science and technology studies in order to assess their critical potential for a future history of journalism studied through the lens of technology.

The Culture of Speed: Telegraphic Communication

To start with, the example of the telegraph might be illustrative. Many popular writings on electrical telegraphy emphasize the dramatic increase of speed in the point-to-point transmission of information as the main and most important impact of this new technology. Yet these popular accounts tend to take the revolutionary rhetoric that accompanied the emergence of telegraphy (just as any other new communication technology thereafter) at face value, ignoring the many technical and practical problems and challenges that characterized the use of this new infrastructure on an international, national and regional level. In fact, communicating by electrical telegraph was a highly formalised procedure that demanded a lot of tacit knowledge by its operators (Laborie 2010). The transmitting and receiving parties were engaged in a series of reciprocal actions aimed at securing trouble-free transmission and controlling the communication act. In this sense, the electrical telegraph was not just a technology for the transmission of signals at the speed of electricity. As a new medium, it also affected the form and style of communication at a distance. The "telegram style" became a new genre of communication, affecting not only long-established forms of diplomatic correspondence but also the production and presentation of "news" in newspapers and gazettes. Like new transmission technologies in the digital age[2], such as e-mail, SMS, or Twitter-posts, the telegram introduced a new form of writing. In concentrating on the essence of a message, and favouring a

2 This "culture of speed" as John Tomlinson (2007) has coined it has by no way reached its end. As the amazing success of the Huffington Post shows, the raise for outspeeding competitors on the news market by using new online services such as Twitter or Facebook as dissemination platforms deeply affects editorial and journalistic practices. Hamann, Götz (2013). Alle 56 Sekunden was Neues. *Die Zeit*, October 2, 24.

short, fact-oriented narrative, telegram style symbolised a new kind of "rationalized" communication.

Information has always been a commodity, but the introduction of the telegraph probably showed at its most dramatic the mutual relationship between time and money. The acceleration of the transmission of messages led to the rationalisation and acceleration of the production (writing) and dissemination of messages, which in turn affected the established networks for the circulation of "news," the newspapers and gazettes. As Jürgen Wilke (2004) and Menahem Blondheim (2004) have shown, the "imperative of condensation" had measurable impact on more expressive and stylised messages as well. Despite the qualitative shortcomings in electrical communication at a distance, the "cult of speed" paved the way for an ever-increasing acceleration of the production and dissemination of "news" (Tomlinson 2007). The speeding up of the electrical dissemination of news coincided with a boost in the rationalisation of production technologies in printing. These parallel processes of acceleration and rationalisation in the production and dissemination of news and its mediated circulation in newspapers and journals catalysed the spread of so-called "dailies" from the 1850s on.

But the development and implementation of new technologies for the transmission and production of news alone cannot explain the "success" of the daily newspaper business. Equally important was the emergence of a popular press culture, focusing on human interest, crime and spectacular news. In keeping with this formula, a new editorial policy was launched, with short articles written in a simple style. The dailies were targeted mainly at a new reading public in the lower middle-class and the educated working class. The trend to short, direct items (that is, in telegram style!) allowed the papers to be scanned by hard-working people who didn't have much time for reading. Nevertheless, readers got the feeling that they themselves could select what they deemed interesting or important. As Donald Sassoon has put it: "Scanning the paper empowered the reader" (Sassoon 2006, 329). The acceleration of news production and dissemination thus coincided with a new form of news consumption, new laws concerning the freedom of expression and the emergence of specialised news broker, such as the news agencies (Shrivastava 2007).[3] Mutually, these developments generated a dialectic

3 The speeding up of news consumption coincided with other forms of accelerated consumption—for example of alcohol. As Wolfgang Schivelbusch (1992) has shown, industrialization in Britain and its effects on the working class led to a sensitive shift in drinking habits. The beer culture, characterized by the relatively time-consuming consumption of

process of ever faster production and demand for 'news' characteristic of the mass media market ever since.

From the point of view of the history of technology, the apparatus that best symbolises the accelerated dissemination of news is the ticker. The news ticker was based on the principle of the stock ticker, invented by the American Edward Calahan in 1863. The receiving end of the stock ticker was a printing mechanism that produced the distinctive sound that gave the system its name. The ticker enabled the electrical transmission of quotes for stocks, bonds and commodities directly from exchange floors to a multitude of reception devices owned by brokers and investors across the country. The New York Stock Exchange (NYSE) was an early adopter in 1867. In 1871, Thomas Edison designed an improved version of the device that he called a 'Universal stock ticker' (Israel 2013, 33). By 1874, he had sold more than 5,000 of them. The synchronisation of two realms—a centralised production and dissemination process of information (be it stock prices or news) and the multiple and decentralised reception of information—created economies of scale that allowed for an enlargement of the scope of the new electrical information infrastructure.

This infrastructure, although controlled or regulated by different institutional actors (nationalised telegraph services in many European countries or private networks), produced some unintended consequences though. In a comparative analysis of the impact of the new telegraphic infrastructure on two Swedish newspapers, the *Norrebottens-Posten* in Piteå (in the north-east of Sweden close to the Finnish border) and the *Öresunds-Posten* in Helsingborg (in the south-west of Sweden close to Denmark), media historian Jonas Harvard was able to draw a number of interesting conclusions about the spatial and temporal impact of the new technology on news reporting (Harvard 2007). Both cities were connected at an early stage to the nationalised telegraph network, Helsingborg in 1854 and Piteå in 1857. By the end of the 1850s, Sweden was linked to Denmark by a submarine cable, laid in 1855, and through it to Hamburg, which was connected to the Russian network via Harparanda and the Finnish lines. But while the *Öresunds-Posten* embraced the new technology as a means to further strengthen its position as a mediator of news heading north via Copenhagen and Hamburg, *Norrebot-*

large quantities of low alcoholic liquid, was challenged by the gin culture, reflecting the tendency in the working class to get drunk faster by the consumption of highly alcoholic spirits.

tens-Posten reacted more conservatively.[4] In part, this was undoubtedly due to the fact that Piteå is smaller than Helsingborg, with a less active economy. That circumstance was also reflected in the frequency of publishing—weekly in Piteå and thrice-weekly in Helsingborg. But according to Harvard, the differences in the adoption of the new technology were due less to social and political determinants than to geography and climate. Harsh winters and endemic vandalism severely affected the reliability of the telegraph lines to the north of Sweden. The high cost of employing a correspondent in Stockholm to supply the small local newspaper with news from the Swedish and other European capitals seemed excessive, given the limited interest in these matters on the part of the readers of the *Norrebottens-Posten.* It was sufficient to report on the main European news stories with articles taken from the national newspapers. Although these would arrive with some delay by ordinary mail to Piteå, the *Norrebottens-Posten* did not scruple to give them the byline "telegraphic news" or "the latest telegrams." Only when it came to extraordinary events—mainly political crises or wars—was the telegraphic infrastructure really employed to provide readers in the vast territories of a country like Sweden with speedy coverage.

In day-to-day reporting, it was acceptable (even for the *Öresunds-Posten*) to stick to older, well-established channels of news transmission via regular mail and business networks, even if that meant a delay of three to four days. As Harvard observes, it was apparent to all that news reporting had several levels of urgency. The initial effect of telegraphy was to reinforce the spatial and temporal divide between the peasant regions of the north and the economic and cultural centres of the south. "When the telegraph worked as it should, in the 1870s *Öresunds-Posten* could get news *faster* from cities far away then from nearby locations in the province. The telegraph placed news from Berlin, Paris, London and Vienna on a single temporal scale, and regional news was left behind" (Harvard 2007, 80). The interesting findings of Jonas Harvard thus force us to rethink the popular narrative of the electrical telegraph as the "Victorian Internet" (Standage 2007) and remind us that the business of news gathering and dissemination was largely local and regional. The electric telegraph, although almost always interpreted as a new

4 While the use of the telegraph as news transmitter remained rather limited in Piteå (reaching approximatively 1,000 telegrams in 1862), the number of telegrams passing the telegraph office of the *Öresunds-Posten* amounted to almost 50,000 in the same year. On December 2, the *Öresunds-Posten* reported that the telegraph personnel had become sick from the intensity of their work!

medium with the power to annihilate time and space, also had contrary effects. It was capable of fragmenting communication spaces and establishing territorial friction between privileged and underprivileged gateways or nodes of electrical transmission and reception of information. This tension both predates and post-dates the advent of the telegraph and must rather be characterized as a stable historical pattern in the development and use of large technological infrastructures in communication.

The "Electrical Sublime" of Wireless Technology

The emergence of wireless technology some fifty years later might serve as another example of a process of socio-technical co-construction, but I would like to use this case in order to demonstrate the importance of the symbolic capital attached to every new technology, especially in the field of telecommunications. According to Charles Bazerman, the author of the fascinating study on *The Languages of Edison's Light*, a technology must not only succeed materially; it must also succeed symbolically. "In building enduring meaning and values for the technology they wish to implant in our daily lives," he argues, "the creators of that technology must engage in heterogeneous symbolic engineering" (Bazermann 1999, 335). But as the example of wireless telegraphy and radio shows, this investment into the symbolic capital of technology is a process that involves many actors of a society and often lies beyond the control of inventors and entrepreneurs. The emergence of wireless technology at the turn of the 20th century brought about a tremendous revival of the 'electrical sublime' that had characterized the early days of wired electrical telegraphy (Marvin 1998; Nye 1996). This revival took place in popular as well as scientific literature. The imaginary dimension of wireless was related to the fact that people could experience the technology in operation without understanding the physical principles of its functioning. The experience of speaking into the air—through a mysterious carrier medium called "the ether"—was loaded with religious associations and often accompanied by spiritual speculation (Peters 2001). The fascination of radial communication—in contrast to the linear transmission of messages along wires—was grounded in the possibility of transmitting a message to an unknown number of potential listeners or receivers at the same time. The combination of global reach and simultaneity fired an imaginary perception of the world

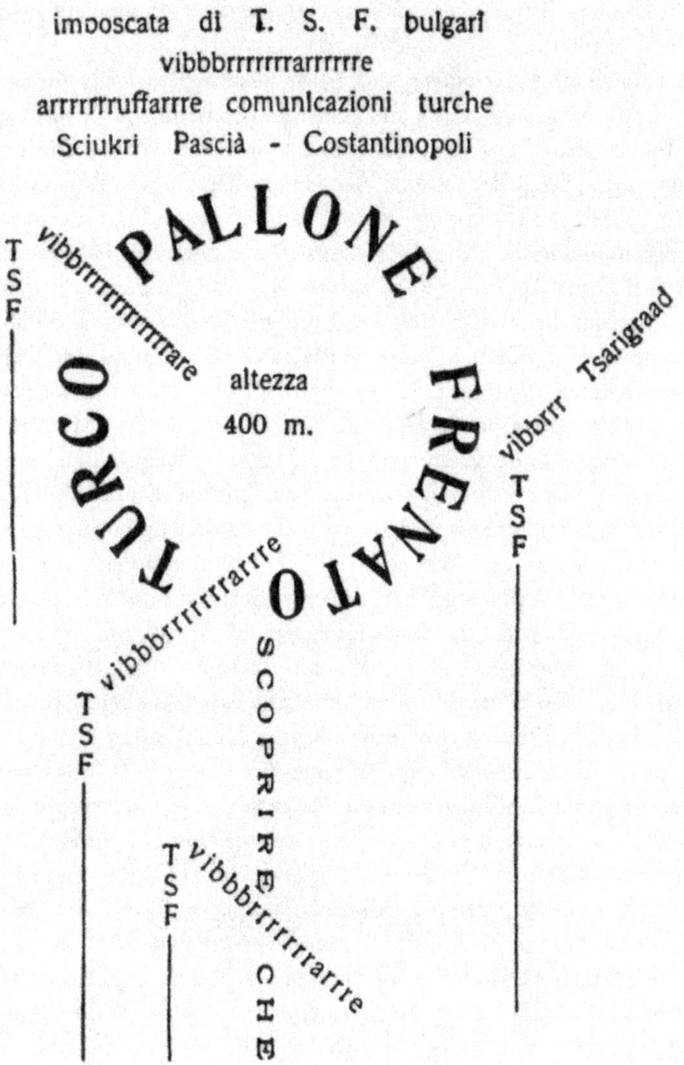

Fig. 1: Marinetti, Zang Tumb Tuuum, Adrianopoli Ottobere 1912 (excerpt)
(Source: Daniels 2002, 123)

that found its artistic expression in a new "poetics of simultaneity" (Kahn & Whitehead 1994).

Radio amateurs, scientists and artists alike experienced wireless communication technologies as a modernist and sometime avant-gardist gateway for the spatial and temporal connection to the world (Sconce 2007). Around 1913 a lively debate took shape in art magazines in Paris, Milan and Berlin concerning priority in the invention of the new artistic movement called "Simultaneism." The main actors in this discourse were the Frenchmen Guillaume Apollinaire, the painter Robert Delaunay, and the writer Henri Barzun; the Swiss-born novelist Frédéric Louis Sauser; and the Italian protagonists of the Futurist movement Umberto Boccioni, Carlo Carrà and FilippoTommaso Marinetti. The publication of the '*Manifeste du futurisme*' in the French newspaper *Le Figaro* on February 20, 1909 by Marinetti initiated a European-wide debate amongst artists from various disciplines about the role of modern technology in society and culture (Daniels 2002).

But it was thanks to his activity as war correspondent for the French press during the Italo-Turkish War of 1911–12 that Marinetti made close contact with the wireless technology that so deeply affected his attempt to renew the language of poetry. He translated his experience of '*telegrafia senza fili*' into an '*immaginazione senza fili,*' systematised in the formulation of the theory of '*parole in libertà*' (words in freedom). His eyewitnessing of the siege of Adrianopolis in 1912 led two years later to the publication of the 21-page-long poem '*Zang Tumb Tuuum, Adrianopoli Ottobre 1912.*' In the poem he recapitulated the hostilities in a new telegraphic style, without punctuation or syntax and organised in a simultaneous presentation of noises, letters and numbers spread all over the sheet (Campbell 2006, 67–96; Campbell 2009).

In a very similar style of "freed words," Guillaume Apollinaire published the '*Lettre Océan*' in his journal *Soirées de Paris* in June 1914. Yet the German art historian Dieter Daniels draws an important distinction between the two creations. He interprets Marinetti's '*Zang Tumb Tuuum*' as an artistic appropriation of the telegram style based on Morse code, while he sees Apollinaire's '*Lettre Océan*' as a poetic translation of the new possibilities of wireless broadcasting. The technique to transmit sound and music led Apollinaire to invent a new form of poetry, inspired by the imaginary of radial communication (Daniels 2002, 124–129).

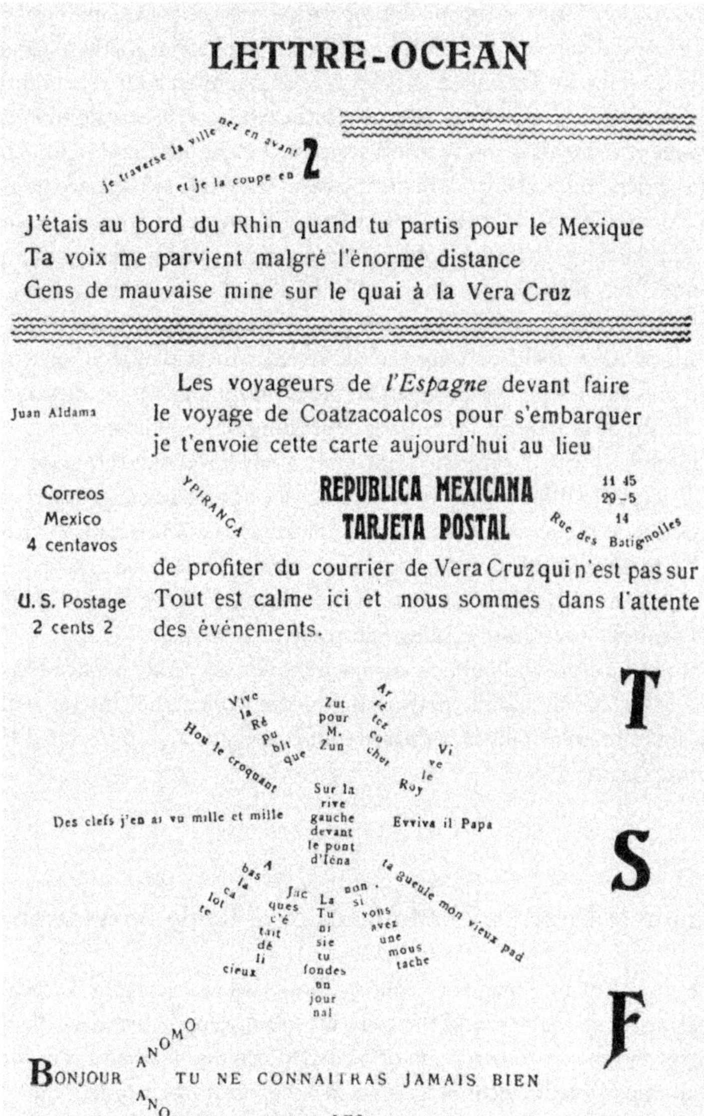

LETTRE-OCEAN

je traverse la ville nez en avant et je la coupe en 2

J'étais au bord du Rhin quand tu partis pour le Mexique
Ta voix me parvient malgré l'énorme distance
Gens de mauvaise mine sur le quai à la Vera Cruz

Juan Aldama

Les voyageurs de *l'Espagne* devant faire
le voyage de Coatzacoalcos pour s'embarquer
je t'envoie cette carte aujourd'hui au lieu

Correos
Mexico
4 centavos

YPIRANGA

REPUBLICA MEXICANA
TARJETA POSTAL

11 45
29-5
14

Rue des Batignolles

de profiter du courrier de Vera Cruz qui n'est pas sur
Tout est calme ici et nous sommes dans l'attente
des événements.

U. S. Postage
2 cents 2

Vive la Ré pu bli que
Zut pour M. Zun
Ar rê tez Co chet
Hou le croquant
Vi ve le Roy
Sur la rive gauche devant le pont d'Iéna
Des clefs j'en ai vu mille et mille
Evviva il Papa
A bas la ca lot te
Jac ques c'é tait dé li cieux
La Tu ni sie tu fondes un jour nal
non si vous avez une mous tache
ta gueule mon vieux pad

T S F

BONJOUR ANOMO ANORA TU NE CONNAITRAS JAMAIS BIEN LES

Mayas

Fig. 2: Apollinaire, Lettre Océan, 1914
(Source: Daniels 2002, 266–267)

This excursion into the artistic appropriation of wireless technology is important, as it demonstrates a shift in the intellectual appropriation *of* and the symbolic connotations linked *to* electrical telegraphy in a long-term historical perspective. The early discourse on electrical telegraphy was dominated by amazement at the new ability to send specific bits of information like stock-market prices, news and diplomatic messages from one specific geographical point to another at the speed of electricity. Discourse on wireless technologies, in particular wireless telephony, was characterised by a fascination with *how* and from *where* information could be spread and received. The linear logic of telegraphic transmission gave rise to discussion about *what* and *how fast* information could be transmitted, whereas the radial logic of wireless telegraphy and telephony triggered fantasies on *from where* the message was sent. With the emergence of radio broadcasting, interest in *what* was transmitted merged with curiosity to know *from where* it was sent (Douglas 1999).

Or, to put it differently, with the emergence of broadcasting, a new quality occurred in the process of the world's spatial and communicative networking. From a macro-historical perspective, one could describe this process as a change from a visible to an invisible networking of the world. While all kinds of traffic routes and cabling (electricity, telephone, telegraph) left visible traces in private and public spaces, wireless broadcasting heralded the age of the invisible but audible networking of the world. This process resulted in an increased sensibility for the sense of hearing as a producer of cultural meaning (Sterne 2003).

Medium or Message? The Immediacy of Radio Broadcasting

The example of the imaginary spill over of wireless telegraphy is speaking when one tries to understand the complex relationship between media technologies and the co-construction of mediated realities. It sensitizes us for the dual nature of media technologies when it comes to their function as both mirror and actors of our life and it forces us to rethink the role of media technologies in the production, dissemination and appropriation of mediated messages. While the avant-gardes enthusiasm for wireless telegraphy and telephony could be interpreted as a classic example of what historians of technology Mikael Hård and Andrew Jamison (1998) have coined the "intellectual appropriation of technology," the emergence of radio broadcasting could be described as a bottom-up process of appropriation of a new communication technology.

The early phase of radio technology was clearly driven by a mass of so-called radio amateurs or "hams," who were fascinated by the possibility of both receiving and sending messages through the ether (Haring 2007). In promoting the tinkering with technology as a basically male and juvenile activity and exclusive expression of technological modernism, radio amateurs—often organized in radio clubs—practiced a new form of technologically mediated communication that deeply altered traditional forms of face-to-face communication (Peters 2001). As Susan Douglas and others have shown, the act of listening in early broadcasting history was by no means a passive exercise of mass media consumption, but an active form of "explorative listening" that involved both technical and social skills by the radio operators (Douglas 1999, 55–82). Although the advent of broadcasting meant an important shift in the communication model turning the two-way model of wireless telephony into a one-way model of radio broadcasting, the radio users remained active participants in the construction of a newly mediated reality (Schrage 2001).

But the example of radio is not only relevant because it demonstrates the co-construction of mediated realities by new communication technologies and its users, but, as mentioned earlier, it asks for a more bottom-up approach when it comes to the material appropriation of new media technologies (Fickers 2012a). Individual station-adjustment (fine-tuning) was of central importance in the first decade of radio broadcasting, that is to say, until the emergence of calibrated station scales in the late 1920s. The fine-tuning adjustment asked for a concentrated dedication of haptic and acoustic senses, or great "fine motor skills" (Schmidt 1998, 316). It is impossible to underestimate the tacit knowledge needed to operate a receiver. To reconnoiter the broadcast spectrum, many circuits had to be adjusted individually by rotating the control knobs sequentially in small increments until a station was heard. Next, several precise readjustments were necessary to bring in the desired program with clarity and volume (Harrison 1998, 296). The gradual development of automatic tuning devices, then, was an indispensable condition for the radio set to transform from a tinkerer's or amateur's medium to a mass medium. Central steps in this process were the embedding of previously exposed components into a cabinet, the reduction of tuning knobs, and the integration of the loudspeaker into the cabinet. From 1924 onwards, there were over seventy patent applications for single-control tuning in the

US alone (Harrison 1998, 299–300), while the radio industry sought to produce sets for everyone. They reckoned that radio listening should become dead easy and the technological vision of time was a receiver that could be operated with only one knob.

As Roger Silverstone and Leslie Haddon have argued on the basis of Adrian Forty's study of the English radio cabinet design between 1928 and 1993 (Forty 1972), the gradual simplification of the operation of the radio set since the mid 1920s is a fitting example of what they call "pre-domestication": "an anticipation in design itself of the artefacts' likely place (in this case the home) and an attempt to offer a solution *in the design of the object itself* to the contradictions generated within the process of technical innovation" (Silverstone and Haddon 1996, 49). Because of the "double articulation" of the radio as both object *and* medium—a key distinctiveness of media technologies compared to other technologies—Haddon and Silverstone speak of successful domestication only when a communication technology has been tamed as an object and socially appropriated as a medium. "With both dimensions of doubling in mind it becomes both possible, indeed essential, to see the innovation of media and information and communication technologies as a fundamentally social process, and the particular complexity of the design/domestication interface of these technologies as one which requires an understanding of the role of both producer and user in its definition" (Silverstone and Haddon 1996, 50).

Within less than ten years, the radio became domesticated both as an apparatus and a medium. Of course, this process was not an uncontested one and showed—from a retrospective point of view—all typical symptoms of the discourse of ambiguity or "binary thinking" that characterize the emergence of new media (Sturken and Thomas, 2). While the public debate about new media is an important factor in the creation of a horizon of expectation in which each new technology is discursively embedded (Boddy 2004; Goody 2011), another factor seems pivotal when analyzing successful processes of media innovations: the capacity of a new medium to incorporate and reconfigure existing norms or logics of media aesthetics and modes of audience address. The example of television might serve as a prime example of such a process of remediation (Bolter and Grusin 2000), in which established reality codes in audio-visual representation of the world stemming from the "old" media of radio and film became successfully incorporated into the business model and institutional framework of television and—after some experimental phase—translated into a new aesthetic of televisual representation of the world.

New Reality Codes of Audio-Visual Representation: Liveness, Authenticity and Immediacy

While most of the recent studies looking at processes of remediation focus on the incorporation and appropriation of existing media products and practices into the new digital media following a logic of convergence (Jenkins 2006), it is important to emphasise that such processes always involve mutual interference and transfer. The rise of television not only changed the fragile balance in the hitherto existing mass media ensemble but forced all media to reposition themselves and to react to the "threat" of the newcomer. The invention, development and implementation of a new communication technology and medium is per definition a phase of uncertainty, of negotiation and of creative exploration (Winston 1998). Steve Wurtzler (2007) has explored the intermedial relationships of interrelated sound technologies of the 1920s and 1930s (the sound cinema, the phonograph and the radio) which developed in close interaction. He highlights the complex and hybrid relationships between technologies as object, practice and discourse: "Upon their innovation, media technologies have multiple, often conflicting identities. The ultimate meanings they take within social relations are the product of contestation and struggle. Various mechanisms—some economic or regulatory, others ideological or material—function to privilege one set of meanings over others as media technology is successfully secured within social habits and practices" (Wurtzler 2007, 15). Instead of emphasising the "revolutionary" character of this process, Wurlitzer shows that the grafting of electrical sound technology onto the existing institutions and practices of cinema and phonograph was "conservative," both in an economic, aesthetic and technological sense. The material and rhetorical struggles to shape the identity of radio as a new medium led to a process of reconsolidation within the existing media landscape— "institutional realignments related to technological change exacerbated ongoing consolidation within each industry" and forged the re-articulation of previously established positions on cultural and political life (Wurtzler 2007, 69).

In a similar vein, I have argued that the emergence of television must be interpreted as conservative media revolution (Fickers 2012b), characterized by: 1) an institutional growth of television within the existing structures of radio broadcasting; 2) the successful incorporation of narrative forms (for example genres) and aesthetic conventions (for example filmic styles) from existing media into the "parasitic" medium of television; 3) the recycling of well established spatial as well temporal metaphors for the description of an apparently new and revolutionary audio-visual experience (for example "window of the world"); and finally 4) the strong emphasis on the nation, the home, and the family that made television a privileged actor of the moral and material recovery of the nation state in post-war societies.

The intermedial perspective clearly helps to relativize the revolutionary rhetoric that dominates both popular and scholarly accounts on media history. But on a more fundamental level, it challenges us to reflect more deeply on the role of technology in the creation of new codes of reality that are inscribed into the materiality of media technologies. In other words to think about the symbolic capital of technology when it comes to the construction of concepts such as "liveness," "immediacy," and last but not least: "authenticity" (Gripsrud 1998). All media technologies discussed so far have not only altered peoples' lives in terms of new forms of electronic production, transmission and reception of mediated messages (be it news, music or serial entertainment formats), but they have changed traditional forms of storytelling and communicative interaction with and within the world by establishing new narrative styles and conventions framing our perception of reality (Siegert 1999).

While the newsticker and the typewriter changed the production, dissemination and presentation of information in newspapers and magazines, the invention of recording technologies both for images (photography or film) and for sound (gramophone or magnetic tape) deeply affected our ways of seeing and hearing (Suisman and Strasser 2010). Because of the high visual and sonic evidence of photographs, moving images, gramophone or magnetic tape recordings and their indexical relationship—at least when we speak about analogue media—to the recorded reality, media theorists and cultural critics have emphasized the privileged status that these technologies were attributed in discussions about authenticity and accuracy of medial representations of reality (Doane 2007).

But this fascination for or criticism against the technical feasibility of mechanical and nowadays digital reproductions of real life sounds and images in popular and scholarly works and debates has probably misdirected our analytical perspective, misdirected in the sense that the discussion of the potential or real impact of such technologies on our social, cultural or political lives has been framed in a dual logic of either good or bad rather than exploring the meaning of the Kranzbergian option of "non-neutrality." What could non-neutrality mean when thinking about communication technologies such as radio, television or the internet? Instead of engaging in the normative debate about wishful or harmful effects of media products and technologies, historians of technology and media could contribute to the analysis of how technology acts as a mediator between form and function, medium and message, or reality and representation. The magnetic tape recorder, the microphone, the transmitter, and the radio receiver all intervene in the complex construction of a radio documentary to no lesser degree than do the narrative plot, the voice-over, and the dramatization of the story by a careful montage of sound bites and music (Hand and Traynor 2001; Birdsall 2013). The "authenticity" of a television programme is always the result of a careful mise-en-scène, playing with the audio-visual evidence of live television and documentary filmmaking, unmasking both as the being the result of specific narrative styles and conventions (Häusermann 2007; Ellis 2011). Both play with the spectator's desire for reality—be it on the "live" mode of television or on the held and reviewable mode of documentary radio and film.

As Elizabeth Cowie (2011) has convincingly argued in her book on documentary filmmaking entitled *Recording Reality, Desiring the Real*, our fascination with the spectacle of "real time" re-viewed is deeply embedded into the modern pleasures of looking, and characterized by the paradox of recorded reality being experienced both as spectacle and information or knowledge. The narrative conventions of radio, television and documentary filmmaking deploy specific modes of storytelling, all relying on their claim to "truth" as non-fiction. These conventions frame our perception of recorded reality, to such a degree that even if the "recorded reality" is faked, we tend to believe what we see because of the visual evidence of moving of images and the authoritative voice-over of a main narrator. Understood as mediating interface between reality and representation, communication technologies are never "neutral" actors in the production, circulation and appropriation of "information" or "messages," but are non-neutral "actants" in the Latourian sense of the term (Latour 2007). Technological actants intervene in the production and reproduction of mediated messages, technology shapes specific formats of storytelling and narrative conventions, and technology frames our perception of reality. Yet technology does not determine the way we experience reality or how we interact with or participate in the world. In a process of co-construction, technology and its users continuously negotiate both the means and meanings of the flows of information.

Technology as *dispositif*

This last remark forces us to reflect upon the implicit relationship between technology, medium, and message—or, in the words of the French philosopher Jean-Louis Baudry, to think about media technologies in terms of their *dispositif* (Baudry 1978). According to Baudry, the materiality of a medium *('l'appareil de base')*, the content that is transmitted, screened or circulated and our perception of this content (texts, images, sounds) in a specific spatial and temporal constellation call for each other. In other words, they are mutually inclusive. In this concept, the materiality of media technologies is logically bound to the social practices and symbolic meanings it co-produces. Technology intrinsically shapes the production, dissemination and perception of our mediated reality, and the narrative and aesthetic conventions that are inscribed into these technologies frame our perception and interpretation of that mediated reality to such a degree, that we can hardly differentiate fiction from faction, representation from reality.

This mutual relationship of the spatial arrangement of technology, narrative conventions of the medium and ritualized forms of public viewing in a specific environment (cinema) defines the classical *dispositif* of film. While Baudry's cinematic *dispositif* is a kind of a-historical or ideal typical theory of the filmic viewing situation and the ideological effect it produces, media historians like Knut Hickethier (1995), Carsten Lenk (1996) or Thomas Steinmaurer (1999) have fruitfully applied the concept of *dispositif* to other media (radio and television) and—more important—historicised it in order pay tribute to the changing character of media technologies and accordingly the varying forms of usages over time. Most media knew different phases of dominant *dispositifs* (they started more or less as immobile, domestic media and subsequently became mobile and ambient) but became increasingly mingled and convergent. Historicising media *dispositifs*—or, in other words, thinking about the life-cycles or biographies of communication technologies—could serve as a heuristic tool for the changing functions of technologies and to investigate the constant renegotiation of their meanings and functions in varying consumption environments.

The main heuristic added value for the study of communication and information technologies therefore seems to be the intrinsic interweaving of the material (technical) basis of media technologies, the spatial and social conditions of interaction with these technologies and the resulting attributions of cultural meaning. The concept of *dispositif* therefore seems to be a fruitful theoretical frame for the analysis of the complex processes of interaction between the different dimensions (material, institutional, social, economic, cultural) of communication technologies.

Conclusion

When Kranzberg argued that technology is neither good, nor bad; nor neutral he quite rightly pointed to the fact that technology's interaction with the social ecology is such that technological developments frequently have environmental, social, and human consequences that go far beyond the immediate purposes of the technical devices and practices themselves (Kranzberg 1986). Although Kranzberg referred to technologies such as nuclear energy, medical technologies or DDT when outlining his program in 1985, his observations apply even more convincingly to the field of media technologies which interact—to use another of his terms—profoundly with "the entire sociocultural milieu."

In the course of this brief journey into the history of some emblematic media technologies we discussed a number of elements or factors that seem relevant when thinking about the role, impact, or function of technology in the history of media and communication. New and old media technologies shape—often before becoming a daily life reality—our cultural imagination, thereby placing potential uses or effects of communication devices or products at the very heart of popular imagination and critical debate. Such discourses, mostly characterized by an ambivalent interpretation about the benefits or dangers of the new device or medium for future society, create a specific horizon of expectation, which—consciously or not—influence both developers and potential users of new technologies. As the author has demonstrated on the example of the radio dial, the symbolic capital produced by such discourses is even inscribed into the materiality of communication technologies with design as most evident interface of such symbolic inscriptions (Fickers 2012a).

But media and communication technologies not only shape our cultural imagination, they alter—on a very concrete and day-to-day basis—our interaction with people and the world. In affecting both the temporal and spatial dimension of communication, media technologies have changed traditional forms of social communication and offered new ways of mediated interaction with the world. The recording and reproduction of sound and images, the transmission of electrical signals by wire and electromagnetic waves, the domestication of electrical machines into furniture, the interaction with electrical and electronic devices via keyboards or touchscreens, the ear and eye witnessing of local and global events via radio, television and online services, or the real-time communication over long distances via telephone or internet—all these innovations have deeply changed our way of perceiving and interpreting the world we imagine as well as the one we live in. Technologies of seeing and hearing as well as writing and printing have produced new modes of coding and decoding our mediated realities (Hall 1980; De Certeau 1980).

In using or non-using specific technologies at their disposal, journalists and their public play an active role in the co-construction of these mediated realities (Nyre 2008). The use of technologies for the production of journalistic narratives—be it a newspaper editorial, a radio feature, a television documentary or a twitter message—is never a "neutral" affair (Dooley 2007). As we have seen, technologies influence the way a story is told and how the audience might perceive or interpret it. What an interdisciplinary study of the rich and complex history of media technologies based on the concept of *dispositif* can offer is a hermeneutic framework that enables us to interpret the role, function and impact of such technologies and practices in their economic, political, social and cultural dimension beyond the simplistic and normative model of good and bad.

Works Cited

Baudry, Jean-Louis (1978). *L'Effet cinéma.* Paris: Albatros.

Birdsall, Carolyn (2013). Sonic Artefacts. Reality Codes of Urbanity in Early German Radio Documentary. In Karin Bijsterveld (ed.). *Soundscapes of the Urban Past. Staged Sound as Mediated Cultural Heritage*, 129–168. Bielefeld: transcript.

Birkner, Thomas (2012). *Das Selbstgespräch der Zeit. Die Geschichte des Journalismus in Deutschland 1605–1914.* Köln: Herbert von Halem Verlag.

Blondheim, Menahem (1994). *News of the Wires. The Telegraph and the Flow of Public Information in America 1844–1897.* Cambridge, Mass.: Harvard University Press.

Blondheim, Menahem (2004). "Slender Bridges" of Misunderstanding: The Social Legacy of Transatlantic Cable Communications. In Norbert Friztsch and Ursula Lehmkuhl (eds.). *Atlantic Communications: The Media in American and German History from the Seventeenth to the Twentieth Century*, 153–170. Oxford: Berg.

Boddy, William (2004). *New Media and Popular Imagination. Launching Radio, Television, and Digital Media in the United States.* Oxford: Oxford University Press.

Bolter, Jay, and Richard Grusin (2000). *Remediation. Understanding New Media.* Cambridge, Mass.: MIT Press.

Campbell, Timothy (2006). *Wireless Writing in the Age of Marconi.* Minneapolis: University of Minnesota Press.

Campbell, Timothy (2009). Marinetti, Marconista. The Futurist Manifestos and the Emergence of Wireless Writing. In Debra Rae Cohen, Michael Coyle, and Jane Lewty (eds.). *Broadcasting Modernism*, 51–67. Gainesville: University of Florida Press.

Carey, James W. (2009). *Communication as Culture.* London: Routledge.

Certeau, Michel de (1980). *L'Invention du Quotidien. Vol. 1: Arts de Faire*, Paris: Gallimard.

Cockburn, Cynthia (1992). The Circuit of Technology: Gender, Identity and Power. In Roger Silverston and Eric Hirsch (eds.). *Consuming Technologies. Media and Information in Domestic Spaces*, 32–47. London: Routledge.

Cowie, Elizabeth (2011). *Recording Reality, Desiring the Real.* Minneapolis: University of Minnesota Press.

Daniels, Dieter (2002). *Kunst als Sendung: Von der Telegrafie zum Internet.* München: Beck Verlag.

Doane, Mary Anne (2007). The Indexical and the Concept of Medium Specificity. *Differences*, 28 (1), 128–152.

Dooley, Patricia (2007). *The Technology of Journalism. Cultural Agents, Cultural Icons.* Evanston: Northwestern University Press.

Douglas, Susan (1999). *Listening In. Radio and the American Imagination.* Minneapolis: University of Minnesota Press.

Ellis, John (2011). *Documentary: Witness and Self-Revelation.* London: Routledge.

Fickers, Andreas (2012a). Visibly Audible. The Radio Dial as Mediating Interface. In Trevor Pinch and Karin Bijsterveld (eds.). *Oxford Handbook of Sound Studies*, 411–439. Oxford: Oxford University Press.

Fickers, Andreas (2012b). The Emergence of Television as a Conservative Media Revolution: Historicising a Process of Remediation in the Post-War Western European Mass Media Ensemble. *Journal of Modern European History*, 10 (1), 49–75.

Forty, Adrian (1972). Wireless Style: Symbolic Design and the English Radio Cabinet 1928–1933. *Architectural Association* Quarterly, 4, 22–31.

Goody, Alex (2011). *Technology, Literature and Culture*. Cambridge: Polity Press.

Grosswiler, Paul (2013). *Old New Media. From Oral to Virtual Environments*. New York: Peter Lang.

Gripsrud, Jostein (1998). Television, Broadcasting, Flow: Key Metaphors in TV Theory. In Christine Geraghty and David Lusted (eds.). *The Television Studies Book*, 17–32. London: Arnold.

Hall, Stuart (1980). Encoding/Decoding. In Stuart Hall et al. (eds.). *Culture, Media, Language: Working Papers in Cultural Studies, 1972–1979*, 128–138. London: Hutchinson.

Hand, Richard, and Mary Traynor (eds.) (2011). *The Radio Drama Handbook. Audio Drama in Theory and Practice*. New York: Continuum.

Hård, Michael, and Andrew Jamison (eds.) (1998). *The Intellectual Appropriation of Technology. Discourses on Modernity 1900–1939*. Cambridge, Mass.: MIT Press.

Haring, Kristen. *Ham Radio's Technical Culture*. Cambridge, Mass.: MIT Press.

Harrison, Arthur (1998). Single-Control Tuning: An Analysis of an Innovation. *Technology and Culture*, 39 (3), 296–302.

Harvard, Jonas (2007). News Heading North: The Telegraph and the Temporal Representation of 'Europe' in the Newspapers of Northern Scandinavia 1850–1870, Research Paper Presented at the *Tensions of Europe Conference*, Rotterdam, June 7–10, 2007.

Häusermann, Jürg (2007). Zugespieltes Material. Der O-Ton und seine Interpretation. In Harun Haye, Cornelius Reiber, and Nikolaus Wegmann (eds.). *Original/Ton. Zur Mediengeschichte des O-Tons*, 25–49. Konstanz: UVK Verlagsgesellschaft.

Hickethier, Knut (1995). Dispositiv Fernsehen. *Montage AV*, 4, 63–83.

Hughes, Thomas P. (2004). *Human-Built World. How to Think about Technology and Culture*. Chicago: University of Chicago Press.

Israel, Paul (2013). 'Claim the earth': Protecting Edison's Inventions at Home and Abroad. In Stathis Arapostathis and Graham Dutfield (eds.). *Knowledge Management and Intellectual Property. Concepts, Actors and Practices from Past to Present*, 19–43. Cheltenham: Edward Elgar Publishing.

Jenkins, Henry (2006). *Convergence Culture. Where Old and New Media Collide*. New York: New York University Press.

Kahn, Douglas, and Gregory Whitehead (eds.) (1994). *Wireless Imagination. Sound, Radio and the Avant-Garde*. Cambridge Mass.: MIT Press.

Kranzberg, Malvin (1986). Technology and History: "Kranzberg's Laws." *Technology and Culture*, 27 (3), 544–560.

Laborie, Léonard (2010). *L'Europe mise en réseaux: La France et la coopération internationale dans les postes et télécommunications (1850–1950).* Bruxelles: Peter Lang.

Latour, Bruno (2007). *Reassembling the Social. An Introduction to Actor-Network-Theory*. Oxford: Oxford University Press.

Lenk, Carsten (1997). *Die Erscheinung des Rundfunks. Einführung und Nutzung eines neuen Mediums 1923–1932.* Opladen: Westdeutscher Verlag.

Marin, Carolyn (1988). *When Old Technologies Were New. Thinking about Electric Communication in the Late Nineteenth Century*. New York: Oxford University Press.

Nye, David (1996). *American Technological Sublime*. Cambridge, Mass.: MIT Press.

Oudshoorn, Nelly, and Trevor Pinch (eds.) (2005). *How Users Matter: The Co-Construction of Users and Technology*. Cambridge, Mass.: MIT Press.

Nyre, Lars (2008). *Sound Media. From Live Journalism to Music Recording*. London: Routledge.

Peters, John Durham (2001). *Speaking into the Air: A History of the Idea of Communication.* Chicago: Chicago University Press.

Richard, John (2010). *Network Nation. Inventing American Telecommunications.* Cambridge, Mass.: Harvard University Press.

Sassoon, Donald (2006). *The Culture of the Europeans. From 1800 to the Present.* London: Harper Press.

Schivelbusch, Wolfgang (1992). *Tastes of Paradise: A Social History of Spices, Stimulants, and Intoxicants*. New York: Pantheon Books.

Schmidt, Uta C. (1998). Radioaneignung. In Adelheid von Saldern (ed.). *Zuhören und Gehörtwerden. Vol. 2*, 259–368. Tübingen: Edition diskord.

Schrage, Dominik (2001). *Psychotechnik und Radiophonie. Subjektkonstruktionen in artifiziellen Wirklichkeiten 1918–1932.* München: Wilhelm Fink Verlag.

Schudson, Micheal (2011). *The Sociology of News*. New York: Norton.

Schwartz Cowan, Ruth (1987). The Consumption Junction. A Proposal for Research Strategies in the Sociology of Technology. In Wiebe Bijker et al. (eds.). *The Social Construction of Technological Systems*, 261–280. Cambridge, Mass.: MIT Press.

Sconce, Jeffrey (2007). *Electronic Presence from Telegraphy to Television.* Durham, London: Duke University Press.

Shrivastava, K. M. (2007). *News Agencies from Pigeon to Internet.* Elgin: New Dawn Press.

Siegert, Bernhard (1999). *Relays. Literature as an Epoch of the Postal System.* Stanford: Stanford University Press.

Silverstone, Roger (1994). *Television and Everyday Life.* London: Routledge.

Silverstone, Roger, and Leslie Haddon (1996). Design and the Domestication of Information and Communication Technologies: Technical Change and Everyday Life. In Robin Mansell and Roger Silverstone (eds.). *Communication by Design.*

The Politics of Information and Communication Technologies, 44–74. Oxford: Oxford University Press.

Standage, Tom (2007). *The Victorian Internet. The Remarkable Story of the Telegraph and the Nineteenth Century's On-Line Pioneers*. New York: Berkley Books.

Starr, Paul (2004). *The Creation of the Media. Political Origins of Modern Communications*. New York: Basic Books.

Steinmaurer, Thomas (1999). *Tele-Visionen. Zur Theorie und Geschichte des Fernsehempfangs*. Innsbruck: Studien-Verlag.

Sterne, Jonathan (2003). *The Audible Past. Cultural Origins of Sound Recording*. Durham: Duke University Press.

Sturken, Marita, and Douglas Thomas (2004). Technological Visions and the Rhetoric of the New. In Marita Sturken, Douglas Thomas, and Sandra Ball-Rokeach (eds.). *Technological Visions. The Hopes and Fears That Shape New Technologies*, 1–18. Philadelphia: Temple University Press.

Suisman, David, and Susan Strasser (eds.) (2010). *Sound in the Age of Mechanical Reproduction*. Philadelphia: University of Pennsylvania Press.

Tomlinson, John (2007). *The Culture of Speed: The Coming of Immediacy*. Los Angeles: Sage Publications.

Wilke, Jürgen (2004). The Telegraph and Transatlantic Communication Relations. In Norbert Friztsch and Ursula Lehmkuhl (eds.). *Atlantic Communications: The Media in American and German History from the Seventeenth to the Twentieth Century*, 107–134. Oxford: Berg.

Winston, Brian (1998). *Media, Technology and Society. A History: From the Telegraph to the Internet*. London: Routledge.

Wurtzler, Steve (2007). *Electric Sounds. Technological Change and the Rise of Corporate Mass Media*. New York: Columbia University Press.

Temporalities of News: News and Speed from the Early Modern Era to the Present

John Nerone

We all know that news travels faster than it used to, and we all believe that this is a meaningful fact. In this essay, I try to think through just how the acceleration of news has been meaningful. I challenge the common sense notion that acceleration has simply made news better—more accurate, more accessible, and therefore a better restraint on abuses by the powerful. Instead, I argue that speed of transmission is meaningful as an element in a dialog with durability. I note that the forms of news incorporate different temporalities in order to make meaning and then to make change. I argue that acceleration in transmission had the ironic effect of creating bottlenecks in the communication system which allowed news monopolies to arise and claim authority, but that further acceleration has begun to disassemble these bottlenecks, with well-known consequences for traditional news organizations. I conclude by questioning whether authoritative accounts of the news of the day will ever again exercise as much power as they did during the high modern period of journalism in the west.

This essay proceeds by analyzing news forms. It starts with some general observations about the temporalities of news. It then focuses mostly on newspapers, and it identifies generic elements that historians of news media will recognize as familiar and recurring vehicles for the freshest intelligence, things like the eighteenth-century news digest and the investigative report. In each of these explorations of a news form, I identify the ways in which novelty takes its meaning against a more durable background, an assumed map of the way things are. This style of analysis assumes that the forms that news takes convey meaning in a deeper way than the specific content of reports (Barnhurst and Nerone 2001). I hope to show how, behind an appearance of an ever-accelerating pace of change, the news manages to reproduce a stable set of categories within its very syntax. This helps explain one of the familiar contradictions of the generation born digital, who simultaneously believe that they are living through the period of the most rapid and fundamental change in human history, and that nothing will ever change.

Why and How Speed Matters

The point of news for the producers and users of news media is to monitor the world in order to update one's map of how the world works. Although one could argue that anything that's novel that happens in the world is news, in practice only a subset of that is taken up by news media. The threshold a happening must clear to become news varies, depending on the news system and the values it embraces, but in any case the news value of a happening will depend on how it speaks to something more durable, something involving a hegemonic notion of how the world works. Deciding what is news involves a dialog between novelty and durability, in other words.

Particular news items often don't explicitly indicate their implications for the enduring map of the social. Some things go without saying. Crime news implies a consensual notion of good and evil, for instance, but one would hardly expect each police report to point out that murder is bad. If an event implies a change in the order of things, though, the news report will usually point this out. A random murder in a neighborhood considered safe will be treated differently than one in a sketchy part of town, not just because "those people" are less important to news professionals, but also because there is already a hegemonic assumption that specific neighborhoods are unsafe—it goes without saying.

It is therefore always necessary for news to move at or near the speed of social change. In any news system, the usefulness of news will be in documenting ongoing changes in the way things work. But this can mean many things. In some cases, it can mean simply providing updates of generic information as quickly as possible. Think of sports news. Scores, injury reports, contractual developments, and other bits of data that keep a fan or bettor up to speed on established sports leagues and competitions just have to move as quickly as possible. Usually an informed fan can figure out the implications for oneself. And there is a dimension of any aspect of the news environment that will be exactly like sports news—in reporting on the economy, financial information moves just like football scores, and so do election returns in reporting on politics. News professionals and scholars don't spend much time thinking about this dimension. It's infrastructural. And it doesn't represent structural change, but rather reinforces existing structures. As the pipeline of sports information increases its capacity and speed, FIFA grows stronger.

More interesting is the dimension of news culture that deals with conjunctural change. Trends in popular culture, or changes in the alignment of political actors, require more effort for both producers and consumers of news. In this sphere, immediacy can be counterproductive to the main uses of news. The Arab Spring rebellions, for instance, produced a flood of immediate impressions through social media applications, but these distracted attention from deep conjunctural changes, and instead produced an instant impression of "twitter revolutions" that has not served well in the aftermath, either for letting news audiences understand what's going on or in letting news professionals find ways to gain the attention of their publics.

In cases of conjunctural change, the news media participate in the construction of an emergent social reality. The misrecognition of a movement by the news media can have obvious and significant effects, as Todd Gitlin demonstrated in the case of the U.S. opposition to the War in Vietnam (Gitlin 1980). Professionals, like the diplomats of the U.S. State Department, develop their own news system to insulate themselves from the distortions of the news. One of the significant revelations of the Wikileaks publication of confidential State Department cables was the existence of a vast parallel system of journalism, rather more honest and accurate than the public system of journalism. While stewarding their own information, these same professionals cheerfully disseminated misinformation to news organizations in order to drive events in certain directions. It was a scandal that the confidential information showed State Department officials misleading public journalists. This is how governing institutions influence the news to try to control developments in moments of conjunctural change. It assumes that there will be a triangular relationship between the flow of information about the world, the interests of empowered actors, and public awareness or intelligence.

Walter Lippmann analyzed this problem in his book *Public Opinion* almost a century ago (Lippmann 1922). He began his treatment with a parable about an island, one not connected to the transoceanic cable system, in 1914. The island was home to some British, some French, and some German citizens. When a ship arrived after a lull of six weeks carrying the most recent newspapers, inhabitants discovered that the British and French nationals had been officially at war with the Germans on the island for the past month.

Lippmann emphasizes the distortion in the heads of individual citizens, who receive wildly different accounts of the world they live in together. It would also be correct to see this story as a lesson about how the world is divided into different informational spheres, part geographic and part social. The raw facts of space had impeded the flow of information to Lippmann's islanders, placing them in a news ghetto; the result was that they were unaware of changes to the most fundamental facts of their political existence. Technology could fix the problem of geography; by the time Lippmann wrote about their plight, these islanders would have probably begun receiving news by radio. The social problem would remain, however. People are sorted into spheres by nationality (in this case), but also by class, race, ethnicity, gender, and ideology.

Technology's conquest of geographical space has arguably made social space more significant. Information can travel instantaneously through the networks that connect banks, so that the world's financescape becomes a single small space. Meanwhile, people living on opposite sides of Woodburn Avenue in Cincinnati continue to experience news flows divergent enough to produce radically different views of the world.

The very abundance of information enhances informational privilege for elites while disadvantaging ordinary people, who lack the capacity to control their exposure to messages. The critique of the twitter-brained public is actually quite old, going back before Lippmann to at least the caricature of the newsmonger in Addison and Steele's *Tatler* and *Spectator* (Cowan 2004). The problem of information overload might be a problem of the fallibility of ordinary folk's cognition, for Lippmann, or a problem of proper political etiquette for Addison and Steele, but for others it has read as a problem of systematic disempowerment. William Manning, writing in 1798, blamed the wealthy for producing an overabundance of newspapers in order to baffle ordinary people, who would not have the time to sort the truth from the many falsehoods that political operators cynically produced (Manning et al. 1993). Manning thought that a good, thorough, politically correct monthly newspaper would be sufficient to keep the citizenry alert. Ten years later, Thomas Jefferson, the sitting President, argued that one would be better informed by reading no newspapers at all (Jefferson 1807). One might summarize this critique as the dilemma of increasing flows of information leading to decreasing accumulations of shared knowledge. This dilemma is heightened in the digital age.

Taming the flow of information is one of the tasks that professional journalism has claimed. Increasing speed and abundance expand the gap between experts who command knowledge and ordinary folk who follow the news. The proper territory of modern professional journalism is in between the heights of professional knowledge and the fecund bottomland that nurtures popular enthusiasms. The journalist mediates between the temporalities of the weeds and the sequoias.

News generally and journalism specifically should be seen as dialogs between novelty and durability. So the temporality of news is really a mediation between the accelerating temporality of information flows and the relatively sedate temporality of accumulating knowledge. Improvements in media technologies figure as exigencies in this dialog. Advances in speed can only be understood as an element in this larger temporal scheme.

To say this is to challenge a habit in vernacular journalism history. Michael Schudson has pointed out that journalists tend to credential themselves by telling stories of their inaugural "scoop": When, as young reporters, they found themselves in the right place at the right time—at a disaster, or an assassination, or a battle—and managed to relay dramatic and unexpected news before other competing newsfolk. Schudson points out that this kind of temporal triumph is something "journalists care about infinitely more than their readers" (Schudson 1986, 80). He is certainly correct. He also notes that improvements in communication technology have made these "beats" smaller in scope—minutes rather than days or weeks—and certainly less important. It is unclear whether this means that journalism gets better as technologies improve, though journalists tend to think so.

Journalists usually recount the development of their craft in terms of their tools. Talk to an old-timer and you'll get a reminiscence about manual Remington typewriters and dial telephones and hot lead. The undertone of nostalgia in these accounts shouldn't be mistaken as anything but backup harmony to the lead melody of the relentless advance of technology, what people mock in charts of corporate performance as "the march to the northeast corner." Journalism historians sometimes identify this rhetoric of the technological sublime as an aspect of the "Whig" or "Progressive" model of history (Carey 1974; McKerns 1977). This venerable line of thinking was already obvious in mid-nineteenth-century claims that the electric telegraph "annihilated space and time" (Czitrom 1982; Carey 1983; Schivelbusch 1986; Mattelart 1996). Karl Marx referred to the "annihilation of space by time" in the *Grundrisse*, echoing this earlier popular description of the telegraph and changing the construction to make an argument about the circulation of commodities (Marx 1993, 539). And certainly if one is talking about specific forms of information, like market information or sports scores, the march to the northeast corner is true.

The steady advance of efficient transmission is also mappable. Allan Pred's work on city systems in the nineteenth century U.S. presents a geography of communication metropolises and hinterlands, and shows how, over time, larger physical spaces were drawn into quicker communication (Pred 1973). This story, one assumes, will end when the entire universe is a single zone of contemporaneity.

Embedded in this model is a critique of speed in transmission that holds that as the news grew quicker and more plentiful, it, and its consumers, became more superficial and more harried. Speed made it more difficult to think, or, put another way, the speed of communication has exceeded the speed of thought. Barnhurst and Mutz have demonstrated that this critique doesn't reflect the actual content of news stories, which, in addition to growing longer and more complex, have also continually included more and more cues to the past and future (Barnhurst and Mutz 1997). Of course, each journalist giving each story a history doesn't mean that any reader reading dozens of stories a day will be able to assemble a meaningful history that informs all of them. Too many histories cumulatively produce noise.

But there has always been a contrary tendency in communication scholarship. In a tradition that runs from Lewis Mumford and Harold Adams Innis through James Carey and John Durham Peters, technology is understood as an element in the cultural construction of reality. This tradition is mistakenly dismissed as "technological determinism"; rather, these writers see technology as an element in a dialectic producing a historically contingent configuration of space and time. As Menahem Blondheim (2003) has pointed out, the evanescence of oral society in Innis's account does not produce an evanescent culture; on the contrary, in reaction to the ephemerality of the spoken word, the culture erects a set of practices that emphasize the durability of tradition and the cyclical recreation of natural time. Communication technologies are part of the ensemble of factors that produce new geographies. In the work of Innis and Carey, they are a primary factor; in David Harvey's work, they are infrastructural to economic systems, affording flows of capital (Harvey 1989).

One way to make this philosophical position more concrete is to focus on the forms of news. I will explore some specific types of news content, particularly from newspapers, over several centuries. Such generic forms of presentation replicated over long periods of time in diverse news systems indicate not just technological affordances but also persisting uses and production routines for news. Such forms of news embody hegemonic values, and work to construct a social order while seeming to serve individuals updating their own social maps.

This paper examines four generic news situations from different eras of the western news system in order to ask questions about the relationship between news and speed. All four are found in many settings and have produced a broadly shared repertoire of news forms. All point to ways in which journalism, as a set of ideals and values informing and disciplining the professional practice of news, continually wrestles the accelerating flow of information.

Newspapers in Colonial British North America

Histories of U.S. journalism often begin with a dismissive treatment of colonial newspapers. They are characterized as poorly written, lacking journalistic enterprise, limited in circulation, dull and confusing to read, and subservient to authority. Worst of all, the news in them was really old: copied from London newspapers, usually, that had compiled translations of continental newspapers that had in turn compiled accounts from correspondents and official documents, mostly regarding the competitions of the various European powers. Encountering such a compilation of compilations of compilations, journalism historians either dismiss it as useless or offer faint homage to its standing as progenitor.

And yet, the energy that went into producing a colonial newspaper was not trivial, particularly considering the difficulty of obtaining printing materials and patronage. Printers must have had a clear sense of the importance of the enterprise. And the expense of subscribing to a newspaper was also not trivial. Readers must also have had a purpose.

One interesting explanation of the uses of the colonial newspaper, offered by Charles Clark (1994), uses James Carey's distinction between the ritual and transmission models of communication (Carey 2008). Clark suggests that colonial newspapers allowed the bureaucrats and merchants of distant colonies to ritually replicate the experience of a London gentleman browsing the news in a coffeehouse. In spite of the antiquity of the content, then, the newspaper still offered itself as an instrument of cultural identification. Anyone who continues to "ritually" consume printed news in the digital age can have some appreciation for this argument, I think.

But what about the transmission uses? These were considerable. Although news was frequently quite old, and usually copied from London newspapers, it allowed colonials to monitor the emanations of imperial events in the neighborhood. News moved at the same speed as administration, commerce, and warfare.

The first colonial newspaper, the *Boston News-Letter*, is a good example of the uses of the newspaper. It appeared in 1704 and was run by Boston's postmaster, John Campbell. The postal system was a key infrastructure for newspapers in terms of both newsgathering and circulation, so it would not be unusual for a printer to occupy that post. This meant that the men and women who ran newspapers often drew a salary that depended on the favor of the government, and they often proclaimed their quasi-official status. The *News-Letter* included the phrase "By Authority" on its masthead. This meant what it said—Campbell considered his newspaper to be an official and approved source of public information.

The *News-Letter* carried miscellaneous news, but not as miscellaneous as it seems to twenty-first-century readers. The content of the *News-Letter* and other colonial newspapers is and was impenetrable to an uninformed reader. The price of admission to this collection of news is at least a solid understanding of geopolitics; moreover, though this would take a long time to prove, it seems to me that it would have been impossible to gain the background understanding you'd need by simply reading newspapers steadily over a period of months. This is because the newspaper does nothing to help the reader.

Colonial newspapers expected their readers to come from a small class of active men (and maybe a few women). The high threshold of geopolitical literacy they required would have taxed ordinary folk. And they were quite expensive, with a year's subscription costing between twelve and thirty shillings, or upwards of a week's pay for an ordinary worker.

But for the privileged reader colonial newspapers provided a useful service, even though the news in these papers seems random, distant, and out-of-date. A typical issue would feature clippings from European and English newspapers on affairs of state in Europe, and events in the European colonies, all weeks or months old, and only a couple of brief items of recent news from the region. These items would be arranged under datelines, which in practical terms referred to the date and place of publication of the newspaper from which the content was copied. This format is confusing to the modern reader, who expects the newspaper to put the news into a form that implies a coherent and meaningful narrative. Now it is possible, and this is what we presume eighteenth-century readers did, to mentally arrange the items to correspond to a master narrative provided by the broader culture and personal experience—to read the items as a set of bulletin updates on the flow of the War of the Spanish Succession, for instance. Printers, however, did not let that master narrative govern the form of presentation. Instead, the format replicated something quite familiar to the gentlemen readers of newspapers: the experience of browsing newspapers in a coffee shop. There the most recent papers were available, and the regulars would go through them one by one, sampling a kind of overlapping array of assemblages of items of news, that they would then make sense of in discussions with whoever else was in the coffee shop. Clark is correct to point out the ritual uses of this kind of reading experience. But there are also specific practical uses.

To the involved reader, say a Boston merchant, even months-old news would be quite important. The European items would have American implications. Tensions between Britain and Spain affected trade in the Caribbean, where Boston merchants shipped sugar and slaves; tensions between Britain and France translated quickly into warfare on the border with France's colonies in Canada, which also affected relations with Native Americans. As for the age of the news, if the point of an item about tensions in Europe was their effect on colonial affairs, then news traveled at the speed of history. Military commands, for instance, were transmitted by ship to North America at the same speed as printed news traveled to civilians. It might take weeks or months for either to arrive, but they would tend to arrive at about the same time. So three-month-old news from Europe would have been very timely to people who knew how to use it to interpret the way things were going to happen in the Americas.

Newspapers also provided authoritative texts of official proclamations. When the Queen addressed Parliament, or when the governor addressed the assembly, newspapers would publish verbatim (and "by authority") the text of these remarks. Again, this content would not have done much for ordinary folk, but would have been meaningful to the highly involved reader, who would have known how to decode subtle cues and evasive language. For them it would also have supplied raw material for their own discussions and actions.

Finally, the postal system was a peer-to-peer open-ended network for transmission. Any node on this network—any newspaper, for instance—sent and received at the same time. So when an original item appeared in the Boston News-Letter, like election results or the Governor's Address to the Provincial Assembly or an account of an encounter with French privateers, it was assumed that that item would be copied by London newspapers and read by officials at Westminster in the same way as provincial newspapers copied the London papers and were read by provincial officials. In fact, much of the communication of the British Empire's bureaucracy came in the form of copies of local newspapers enclosed in official correspondence. This feedback loop would weaken as speedier communication technologies accelerated news transmission.

Telegraphic News Enters Newspapers

The impact of new technologies always comes from a negotiation with facts on the ground. Inherited forms of news mediated the impact of new communication technologies. The formats of digests of clipped news and of items of editorial reportage that were characteristic of printerly newspapers persisted for centuries and helped structure the forms of presentation of later flows of information, including electronic ones. Existing divisions of labor in news organizations and the material interests of the kinds of institutional actors that produced the new flows of news, like stock exchanges, also inflected the uses of new communication technologies. The career of telegraphic news was a case in point.

There is a standard treatment of the telegraph in journalism histories that emphasizes three points. First, the telegraph was the first separation of communication from transformation. Unlike previous improvements—building better roads, putting steam engines in ocean-going vessels, making postage cheaper—the telegraph is treated not as an incremental advance but as eliminating the category of time altogether and ushering in a new age of simultaneity in which space and time are annihilated. Journalism historians didn't invent this meme. It was part of the rhetoric that surrounded the telegraph's introduction (Czitrom 1982; Standage 1998). Second, the telegraph produced cooperative newsgathering in the form of the Associated Press in the U.S. and Reuters, Havas, and Wolff elsewhere (Rantanen 2006). Because of the aspects of natural monopoly built into telegraphic technology, news services also became monopolistic (DuBoff 1984; Blondheim 1994; John 2010). And third, the expenses and exigencies of telegraphic communication, together with the business interests of wire services, drove the development of efficient writing styles, including the inverted pyramid form, and therefore the development of objectivity (Carey 1983; Mindich 1998).

These points correctly identify key elements of the telegraph's impact on the news environment, but they elide other elements. Put another way, they describe a kind of common sense about the ways in which news changed in the half-century following the adoption of the telegraph, but in attributing the source of these changes to the technology, they draw attention away from other interesting elements shaping the news. For the purposes of this essay, I will focus on the temporality of news, but there is also an interesting history to tell about newsroom labor, about political developments in the U.S. and elsewhere, and about gendered elements of the distinction between "hard" and "soft" news.

The telegraph was important for news, and news was an important source of demand for the telegraph industry. As Richard DuBoff puts it, "The earliest, and best, customers of the telegraph were newspapers, retail and wholesale merchants, brokers and bankers, as well as 'speculators'" (1980, 467). But newspapers used the telegraph really for only specific kinds of news. One special area of importance was business news. DuBoff again: "With few exceptions, the dispatches carried business and financial news directly useful for a clientele increasingly aware of the enormous benefits of instantaneous communication" (1980, 468). The possibility of advantages in speed of transmission of market information led to a kind of arms race in commercial data, exemplified in the emergence of trading in crop futures (Carey 1983) and the development of widespread stock speculation by ordinary people (Hochfelder 2006), and culminating now in the computerization of stock trading, in which machines talk to machines in an accelerating digital competition for advantage measured in fractions of a second (Yates 1986; Zaloom 2006; Preda 2009). A far more modest transformation of sports information can also be linked to the telegraph (Haller 1979). And less obvious, the telegraph allowed for the collation of weather information in real time, and was a key factor in the creation of modern meteorology, including the National Weather Service in the U.S. (Hughes 1980; Hochfelder et al. 1999).

Compared to its impact on these specialized forms of news, the telegraph's impact on general news was relatively muted. For international news, the formation of an entirely telegraphic communication network was many years in the making. For North Americans, the telegraph could speed news from, say, Halifax to New York City, but the European news still had to come to Halifax by steamship until after the Civil War. Even with a telegraphic assist, the main form of international news in the U.S. remained the news digest attributed to the arrival of a steamship. Daily newspapers in the interim continued to feature columns with headlines like "Eight Days Later by the Cambria," consisting of updates of European and world news carried across the Atlantic by ship. This news digest looked remarkably similar to the news columns of eighteenth-century newspapers, by the way.

National news found a readier telegraphic infrastructure, but nestled within a dominant postal-based political communication system. The U.S. news system was dominated for most of the nineteenth century by partisan papers that emphasized the voice of the editor in selecting news from exchange papers, in introducing and arranging the various items—much like a television news anchor—and in maintaining mutually beneficial relations with colleagues in the party as well as select adversaries. Telegraphic news could matter in this system, but only so much. A good example chosen at random from early in the telegraphic era is the *New York Herald* of January 6, 1848, which features on its main news page first an account of dueling mass meetings, one at the Tabernacle promoting Zachary Taylor for the Whig nomination for the Presidency, and another at Tammany Hall supporting William Worth for the Democratic nomination. Such accounts were either the self-generated reports of the meetings—their "press releases," so to speak—or the product of a reporter in attendance. There follows a long editorial column on a debate in the U.S. Senate on movements to end the Mexican War, focusing on a speech by Senator Calhoun of South Carolina, which the *Herald* had printed the day before, and on a speech by Senator Crittenden of Kentucky, which it was printing a summary of in this issue. Both had been transmitted by telegraph, and the *Herald* was characteristically puffish about the importance of their (very expensive) use of the new technology. A telegraphic summary of the debate that included Crittenden's speech sat two columns over. It was not a full transcription; that would come at a later point, when Crittenden's authorized text made its way through the Democratic Party press.

In retrospect the *Herald* fundamentally misread the short- and medium-term impact of the technology when it predicted that it would transform the style of political communication. "Short speeches, pointed and pithy, will tell a thousand times more on the public mind, in the present electric age of the world, than any other mode or system of speaking. We are going back to the time of Demosthenes. The *Herald* and the electric telegraph are working a revolution." (*Highly Important from Washington*, 2) The habitual speech of nineteenth-century politicians was long, too long for a telegraphic transcription, and too extemporaneous to be recorded verbatim. What constituted the gold standard of political reportage was the authorized, extended, and corrected version of a politician's speech, ready for publication as a pamphlet-length argument. Next to these productions of historical significance, the other material was of less value. An editor's paragraphs might be worth clipping and copying if they were particularly pungent. The telegraphic items were important as only an anticipation of a full account from either the speaker himself or from a correspondent on the scene. These would come through the traditional postal network. This wouldn't change until long after 1848.

Lurking under the political news is another game entirely, the enterprise of earning distinction by presenting value-added news. In the postal era, newspapers earned distinction not by being the first with the news, usually, but by adding to the readily available flow of news either a particularly reliable editorial voice (and in this party membership was key) or particularly complete or clever accounts of local affairs. That second style of content was produced by reporters, a new occupation in newspapers in the U.S. in the 1820s and 1830s. Reporters were meant to be fact gatherers, unlike correspondents, the older designation for what we would today call journalists, who wrote longish letters from distant places, and were meant to have a voice and persona. Market competition among newspapers, which increased through the first half of the nineteenth century in Britain and the U.S., made the production of news by reporters more important to success. Chalaby (1996) argues that this is the origin of the modern notion of news.

Telegraphic news produced another arena in which newspapers could claim distinctive content. This is the point of the *Herald's* inflated estimation of its next-day reports from Washington. And in fact it was the market value of telegraphic news that allowed a cartel to form around European news (Blondheim 1994). But this hierarchical model of commoditized news was anathema to the party system's sense of the uses of political news, which was supposed to be horizontal and networked. The party system remained the dominant logic, at least until the 1870s or 1890s, and perhaps this fact explains why it was that the parties did not establish their own telegraphic news services (Nerone 2006).

Most histories of news cite the Crimean War as the first instance in which the telegraph altered the landscape of public intelligence. Some U.S. historians cite the Mexican War as an earlier example. A more interesting moment, in terms of both duration and complexity, was the U.S. Civil War. Throughout the War, which, like most wars, was a major instigator of innovation in news transmission, one can see the interaction of the postal-based political press system with telegraphic news. This was true both for the people who worked in news organizations and from the perspective of the Lincoln Administration.

Different sorts of newsworkers had to balance different priorities in covering the War. Editors continued to want to serve the interests of their partisan networks, even while facing increased demand from their publics for reliable news from the battlefield. Some newspapers were well positioned to gain information about the War—particularly those in border states. Because these were looked to for fresh news, they could, willingly or not, introduce incorrect reports into national circulation. Because some of these were aligned with the Democratic opposition to the Lincoln Administration, their motives were suspect. Some were targeted by the administration for legal punishment; more were threatened by crowd pressure. (Nerone 1994).

The leading national newspapers invested in sending correspondents to travel with the armies. These have become famous in the standard journalism histories as the Bohemian Brigade (Starr 1954). Their dispatches were copied by other newspapers and were seen by both political and military leaders as influential. Correspondents relied in part on the conviviality of the armies they traveled with, and particularly the generals, who controlled access to the telegraph. So the reportage that flowed from the armies carried a combination of vigorous first-hand observation and delicate political negotiation. The most interesting form of correspondence was visual. Sketch artists traveled with the armies along with "verbal" correspondents. They behaved like later photojournalists, on the one hand, proud of their willingness and ability to get close to the action. At the same time, they tended to follow the traditions of visual art, depicting action in stylized fashion and following conventions in portraying the character of military leaders. Unlike photojournalists, whose work has come to be viewed as immediate, both in the sense of being transmitted electronically and in the sense of being an unmediated record of what happened, the pictorial journalism produced by sketch artists and the engravers who translated their work bore the clear marks of their intervening intelligence. And it appeared most prominently in weekly newspapers, like *Harper's* and *Leslie's*, where it self-consciously presented itself as memory or history rather than current affairs. Photographers traveled with the armies as well, of course. Their work was not significant for news purposes. They mostly did portraiture, making their money off of selling their services to soldiers who wanted to send a portrait of themselves home, and then stockpiling a macabre portraiture of the dead for future sale to those who wanted to remember the carnage. As it happened, there was not much of a market for that.

Unlike photography, the telegraphic report found a niche in the environment of the news organization. In the long period from the 1840s to the 1880s, telegraphic reporting remained segregated for the most part from the other streams of content in the newspaper. At first, a digest appeared under a heading like "The Latest by the Telegraph." Gradually, the scope of this column expanded until it acquired its own editor. Then, as already noted, telegraphic information colonized financial news, and then breaking sports news. But the bulk of the most important stuff in the newspaper, the national political news, remained tied to the system of postal exchange (Nerone 2008). The *Herald's* talk of revolutionary changes in news, like the popular meme about the annihilation of space and time, seemed more intelligent in the 1840s, when the telegraph was still new, than in the 1860s, when it had been put in its place. Newsworkers and political leaders both seemed to understand this.

President Lincoln was a creature of the party press but also an early adopter of telegraphic communication. He had paved his path to power and peopled his administration with practitioners of partisan newspapering, showing a keen awareness of the workings of the networks of party and reform newspapers. In fact, he recruited his cabinet officers by writing them long letters of invitation—this was the appropriate way of addressing important political leaders—but sent them telegrams that they would be receiving invitations, along with instructions that they should telegraph their responses (Wheeler 2008). He showed the same ambidexterity in dealing with the national press system.

Anonymous correspondence emanated from Lincoln's White House, most particularly from his personal secretaries Nicolay and Hay, both skillful and experienced players in partisan newspapering (Carwardine 2006, 2–6). Unlike previous presidents, Lincoln did not have an administration-sponsored newspaper—partly out of a sense that he did not need one, and partly out of the loss of potential patronage to support one after the creation of the Government Printing Office in 1860. But he had an unusually capable operation lining up editorial support among leading dailies and an even more effective operation spreading news aligned with the administration's position through correspondents and telegraph services. Beginning in 1862, the Administration took control of the telegraph system as needed, and managed the distribution of news from the war; controlling the flow of information was more effective than trying to oversee editorial expression. This infrastructural strategy allowed the administration to continue to embrace constitutional liberties on a legal level (Carwardine 2006; Blondheim 2002). The unfettered functioning of the partisan press, including even the most apparently disloyal newspapers, allowed Lincoln to claim the legitimate support of a deliberating public; at the same time, his operatives tacitly allowed local mobs (often largely composed of soldiers on leave) to threaten and attack opposition newspapers at key moments (Nerone 1994).

Obviously there is much to say about the unevenness of the news environment and the opportunity it presented for the continual play of editorial representation. For the purposes of this essay, I emphasize the temporal dimension. Instantaneous reports seemed to abound, but they had the reputation of being unreliable. Astute readers were aware of this, but anxious citizens could not help but be whipsawed by conflicting reports. Meanwhile, the Administration, concerned that reports could derange public opinion and spark violence on the scale of the 1863 New York draft riots, made a concerted effort not just to limit the flow of unconfirmed news but to actively contextualize reports, using the tools of traditional correspondence and partisan editorializing, along with the less respectable technique of crowd pressure. Put another way, the Administration spent considerable effort extending and slowing the temporality of the news. In this, they had significant sympathy and assistance from newsworkers and party editors.

Investigative Reporting

Investigative reporting appeared a generation after telegraphic reporting and involved a very different temporality. By its nature, enterprise reporting requires an investment of time, often coupled with the exploitation of other types of investigations, like legal proceedings or the work of government commissions. This type of slowly assembled long-form journalism had a kind of counterpart in the human-interest story so common in tabloid journalism. For both investigative and human interest journalism, exclusivity through enterprise provided the exchange value that speed provided for other kinds of news.

One exemplary archetypical investigative series was William T. Stead's Maiden Tribute of Babylon. The Maiden Tribute series, run over successive issues of the London-based *Pall Mall Gazette* in 1885, caused a sensation with revelations about the extent of prostitution in the Victorian city. It made Stead a journalistic celebrity on two continents, a status he enjoyed well beyond his death in the Titanic disaster in 1912 (Nerone and Barnhurst 2012). Much studied today by historians of gender and sexuality, the series is neglected by journalism historians, especially in the U.S.; it deserves to be considered an archetype of muckraking as a literary form, predating Ida Tarbell's *History of Standard Oil*, Lincoln Steffens' *Shame of the Cities*, and David Graham Phillips' *Treason of the Senate*, the series most celebrated by U.S. journalism historians, by more than a decade, and Henry Demarest Lloyd's *Wealth against Commonwealth*, an earlier indictment of predatory capitalism, as well. Stead knew Lloyd and influenced his expose, which became a foundational text for Populists, Progressive reformers, and muckrakers (Digby-Junger 1996). The Maiden Tribute series established several of the characteristic features of muckraking. It invoked a moral righteousness that mimicked evangelical fervor. It crafted a figure of the heroic investigator for its author. And it explicitly linked itself to a reform program.

The key plot element in muckraking journalism is the reveal. The journalist/investigator unveils a social or political reality that has lurked behind appearances (Hofstadter 1955, 186–197). In terms of temporality, this plot move involves a combination of climactic moments of revelation with the implication of a hidden dimension of corruption of long duration. In the Maiden Tribute series, it worked like this: Stead investigated for months, and managed to prove in a night that for decades Victorian gentlemen had been buying underage virgins for their sexual pleasure.

The durability of the corruption revealed in a muckraking report initially seems to suggest a timelessness rather at odds with the received temporality of news. Stead signaled the antiquity of his chosen social problem in his first paragraph by invoking the ancient Greek story of the tribute of virgins that the Minotaur commanded from Athens. Stead asks a senior police investigator at the beginning of his odyssey whether the rape of young girls is going on in brothels in London, and he replies, "Certainly it is, and it will go on, and you cannot help it, as long as men have money, procuresses are skilful, and women are weak and inexperienced." "The powerful deflower virgins," so to speak: what kind of a lead is that? But the power of the example comes by the contrast with modern, Christian, industrial Britain, in which enlightenment and virtue are assumed to have brought about an age of respect for the gentler sex. And this indicates the mission of journalism: to continually haul these remnants of ancient injustice out of the darkness and into the light, where they can be purified. Muckraking is the journalism of the unfinished project of modernity. And Stead points this out in that same first installment. He hopes for reform now because "The hour of Democracy has struck." The rape and exploitation of the daughters of ordinary people is no longer to be neglected.

Also tilting against timelessness is the story of the investigation itself. Any investigative piece suggests the story by which the journalist uncovers corruption and comes to report it; in this implied story, the journalist is always a heroic figure. Stead puts himself boldly forth as a kind of tragic hero and promises the reader that all he knows is first-hand: "But I speak of that which I do know, not from hearsay or rumor, but of my own personal knowledge." And his knowledge has come at the cost of a wrenching personal journey: "For days and nights it is as if I had suffered the penalties inflicted upon the lost souls in the Moslem hell, for I seemed to have to drink of the purulent matter that flows from the bodies of the damned." Why the "Moslem" hell I can't say; but it is a feature of classical epic poetry that the hero must undertake a journey to the underworld. Now most investigative pieces draw upon other institutionally supported investigations—criminal prosecutions, the routine fact-gathering of regulatory agencies, even academic or scientific scholarship. The conventions of investigative journalism elide that bureaucratic investment, however, and emphasize the journalist's voyage of discovery. Stead is explicit about this. His own reporting was prompted by Parliamentary action, begun with hearings in 1882, that had stalled; he undertook a four-week investigation to update the information and "rouse public attention."

This story of the story has its own temporality. First come Stead's initial discussions with knowledgeable men. Then he pushes into the unknown: "into direct and confidential communication with brothel-keepers in the West and East of London and in the provinces." The report then proceeds as a sequence of conversations with various folk in the business in one way or another—brothel keepers, procuresses, current and former prostitutes, and the occasional policeman or detective. Some of these are given back-stories, some not; they are not given names. The overall effect is similar to the presentation of testimony at a court trial; the witnesses are quoted, as if they are testifying in their own words. They all sound a bit like Stead, by the way.

The only story in this installment that features a named victim, a definite time frame, and a specific part of town in which the action took place comes at the end: It is the purchase of "Lily," a 13-year-old girl, who is bought for L5, certified a virgin by a hired abortionist, and then installed in a room in a house of pleasure for the enjoyment of Mr. Stead, though he confirms this only indirectly by saying that he can vouch for every detail based on his personal experience, including her terrified screams when she realizes that "There's a man in the room!" This set of events, which Stead had arranged to prove to the reader that the white slave trade was an actual thing, had occurred within a few weeks prior to publication. All the other stories in the long account were part of the life experience of one of Stead's interlocutors, and had occurred within a longer time frame of several decades.

Three other installments of the series followed in the next few days. The stories created a sensation, with crowds assembling outside the newspaper's offices every day, and other newspapers in Britain and the rest of the English-speaking world following along. And the sensation produced the desired effect: legal action that raised the age of consent for girls from 13 to 16. It also produced an indictment—of Stead himself, for the crime of procuring Lily. He was convicted and sentenced to jail time, which only increased his celebrity. (The Parliamentary legislation he prompted also resulted in the prosecutions of Oscar Wilde and Alan Turing for "gross indecency").

This is a third temporality: the post-story. For any investigative story to be worthwhile, it must spark action in the world, and it must be action that receives public attention. That action can take many forms: the conviction of a criminal, or the exoneration of one falsely accused; the passage of legislation to correct a social problem; an outpouring of public sentiment; or, what is now the most likely and most routinized recognition of enterprise reporting, winning an award. The post-story gives the story an arc across time that insures its worth.

Stead's reporting achieved post-story success partly by replication through a transatlantic network of newspapers linked by cable. The Maiden Tribute series appeared in whole or in part in newspapers throughout the English-speaking world, and reports on the progress of subsequent legislation and of Stead's trial circulated by telegraph. So even though the acceleration of communication did not play a role in the original reporting, it was fundamental to the impact of the story. Had it people not read it simultaneously within Britain and on the far side of the Atlantic, it is unlikely that public opinion could have been recognized as outraged on a scale large enough to produce quick legislation and long-enduring cultural ramifications.

Today's investigative reporters would recognize some of Stead's techniques but reject the series as a model: too many unnamed sources; too many anonymous victims and vague occurrences; way too much moralizing; too much of the personal identity of the reporter. But it is not my intent to say that Stead is the progenitor of modern investigative journalism; only that his series exemplifies and helped to establish certain generic qualities of a successful investigative series. Though it is beyond the scope of this paper, I can suggest that the most celebrated twentieth-century investigative series, Woodward and Bernstein's Watergate reporting, shares many of the same qualities, particularly in regard to the long temporal anchoring, the adventurous story of the story (the reporters wrote *All the President's Men* almost as a treatment for a screenplay, after discussing a movie deal with Robert Redford, in fact), and a continually expanding post-story that includes the trite habit of naming every scandal with a -gate suffix (Schudson 1992).

The temporality of investigative journalism depends in part on the temporality of the telegraph even while it draws on the temporality of classical genres like epic poetry and on the temporality of political history. There is an indirect connection in the political economy of the news between the telegraph and the rise of reporting. Both are ways of commoditizing news.

In the postal age, when printers compiled news content out of found items in other newspapers, official documents, and correspondence, they did not in any meaningful way "own" the news they published. In fact, they produced their newspapers with the hope that other printers would copy their content. Being copied meant being influential and noticed, and it meant that other printers would be anxious to "exchange" which translated into a regular rich flow of newspapers and therefore information into the print office. Competition for news content advanced '*pari passu*' with investments in newsgathering techniques, including hiring postriders and speedboats and using carrier pigeons and visual telegraphs. But only with the electric telegraph did it become normal for a news organization to claim exclusive ownership of national and international news content. In the decade prior to the development of the telegraph, however, a sense of property in local news content had appeared with the hiring of reporters. These early reporters were more stenographers than anything else, sent out into the city to transcribe things like prices in the marketplace, the arrivals of ships and hotel guests, the outcomes of court proceedings, and the official minutes of public meetings. Because newspapers had invested wages in the procuring of this content, they began to express resentment and annoyance when competitors copied it in the same news cycle—that is, when the *Bugle's* court report appeared verbatim by the same morning's *Blunderbuss*. One could argue that this explains the fetishization of the scoop.

The Modern Daily Newspaper Encounters First Broadcasting and then the Internet

The telegraph was the first of many electronic news technologies. Telephones also found their way into newsrooms, in the same generation of technology that produced the linotype and the typewriter. Like the telegraph, the telephone was incorporated into newsgathering practices, but unlike the telegraph, it mainly served local reporting, and it did not produce anything like a wire service. Broadcasting hit the news industry in an entirely different way, producing radio and then television stations that competed with existing news organizations like newspapers for both content, audience, and advertising revenue. Newspapers saw the advent of broadcasting as a direct challenge to their business model. Radio news, it was thought, would steal away the audience of newspapers—why buy a newspaper when you knew all the headlines already from radio? And radio would simultaneously steal advertising revenue, a concern that became much more pressing with the global depression that followed the 1929 stock market crash. In U.S. media history, this moment is remembered as the "press-radio war" (McChesney 1991; Jackaway 1995). It climaxed in the years 1933–1935, as newspapers tried to enforce an embargo on wire service news for radio.

The war ended before very long. A settlement was achieved, in which radio took over the function of providing bulletins of breaking news on the one hand and live coverage of happenings like sporting events or important political speeches on the other. Print news became more explanatory, providing background and expertise that extended a report over a longer temporal trajectory, linking it with past events and projecting its implications for the future. This division of labor in journalism mirrored a division of labor in advertising. Radio became especially good at certain kinds of national product advertising, for instance, while newspapers remained the primary medium for display ads for retailers like grocery stores and department stores. Of course, achieving a stable division of labor was facilitated by the fact that many key radio stations were owned by newspapers.

The rise of television added a complication to this relationship. The visual element of television news seemed to pose a more direct threat to the authority of newspapers on both the national and local levels. Nationally, a small set of broadcasters fielded personalities who quickly gained loyal followings. Locally, conditions varied from country to country, but in the U.S., broadcast stations invested in newsrooms in order to satisfy regulatory requirements for public service, so that in any city of even middling size, three or more television newsrooms fielded twice-daily shows. Meanwhile, in the years following World War II, local newspaper markets had acquired characteristics of what economists call "natural monopoly," in part through the workings of wire services. Why, after all, does any market need two newspapers to carry the same Associated Press reports? Again, then, the newspaper industry viewed broadcasting as poaching on its turf, and attributed a gradual erosion of readerships to television. But this reaction was disingenuous. Newspapers had begun refocusing their efforts on their most lucrative readerships, abandoning working-class readers, for instance, by declining to invest in reporting on labor issues and other matters of concern (Leonard 1995). They worked hard to retain upscale readers, who were more attractive to advertisers, by investing instead in lifestyle reporting (Nadler 2012). In the U.S., the most famous attempt by the news industry to counter the influence of television news, *USA Today*, was also the most misleading. Although it famously emphasized color and charts, and was distributed in newsboxes that were designed to resemble television sets, its design was actually neo-Victorian, emphasizing variety and miscellany. Moreover, it was produced for national circulation rather than local consumption. Most local papers, enjoying monopoly status, were content to lose big segments of readership while continuing to produce profit margins of 20 percent and higher for the rest of the twentieth century.

The impact of the Internet has been more disturbing for newspapers and television news divisions alike. Rather than appearing initially as an invasion and then reaching a stable division of labor, the arrival of digital technology at first seemed pre-domesticated. Many of the affordances of the Internet were incorporated directly into production routines in "legacy" media, and many of the applications of computer technology were developed with newspaper uses in mind (Grier 2013). Only later did news organizations react to digital news as a direct attack on their business model.

With the popularization of web browsers in the 1990s, it became apparent that the Internet would become a widely available point of access to breaking news. At first only early adopters, aficionados, and specialists used Internet sources as a replacement for traditional news delivery, but far sighted observers could see that the bottleneck on authoritative information that daily newspapers enjoyed and had come to share with broadcast outlets would soon be modified, at least, or eliminated. Anyone with a laptop, and then anyone with a smart phone, could set up a news medium instantly and ad hoc: digital technology allowed "mass self-communication," as Manuel Castells observed (Castells 2009).

The arrival of new sites of journalism on the Internet meant traditional news organizations would routinely be scooped by amateurs. Legacy newsrooms, whose personnel still sought to abide by disciplines of verifying, balancing, and contextualizing, could not reliably beat the publication of breaking news by vernacular sources. In many cases, they were chasing press releases, which, instead of going to newsrooms by fax, went directly to the public by Internet, and then by Facebook and Twitter. So again, professional journalism has had to redefine itself around a new division of labor.

As when print confronted broadcasting, professional journalism has tried to distinguish itself from digital media by emphasizing expert observation and explanation, both on the level of the individual story and on the level of the map of the day's events. This means that they respond to the instantaneity and interactivity of digital media by promoting the longer attention spans and deeper wisdom of their news personnel, not just their reporting staff but also the editors who arrange the pages into coherent representations of the world (Barnhurst and Nerone 2001). So far this has been a losing battle, largely because of the significant head start that legacy news had ceded by abandoning its working-class audiences. The mass audience had been one of the initial enabling conditions which, along with control of the bottleneck of information flow afforded by telegraphic communication, allowed news organizations to claim the authority to provide a hegemonic map of "the news of the day."

Conclusion

The news environment has been conspicuously segregated into different temporalities ever since. But this segregation only makes visible what had long been the constitutive tension of news in general and journalism in particular, namely the dialog between the rush of events and a more durable sense of the way things are. News has always had ways of formalizing that dialog, among which are included the forms of news digests, the division of news into categories, the division of labor within news organizations between different streams of content and between various news domains, and the literary structure of news reports, including investigative series.

In an emerging environment of "mobile privatization" (Williams 1972), when technologies develop along a vector of atomized moving individuals encapsulated materially in automobiles and virtually in walkmans and iphones (Hay and Packer 2004), it falls to the dinosaurs of the legacy media—the BBC, the New York Times—to provide a representation of reality that everyone is supposed to acknowledge as authoritative and legitimate. Their editorial energies are invested in this enterprise even as their business managers try and fail to find new ways to propertize it. My confidence that the social order needs such a hegemonic map of the news and therefore will find a way to support it has been weakened in recent years by the incompetence of the leaders of these organizations but strengthened by the existence of new and vast pools of capital ready to be deployed by the firms that have clustered at the new bottlenecks of the digital media system—software manufacturers like Microsoft, search engines like Google, retailers like Amazon, platforms like eBay, hardware manufacturers like Samsung. Many of these are showing an interest in investing in news organizations.

In whatever shape it assumes, the tier of prestige news organizations will find themselves working one step back from the frontline of news. Information will continue to reach the public through quicker and more direct channels, and users will still be able to choose sources which will serve their immediate needs and pleasures. Acceleration assisted the rise of authoritative news only up to a tipping point in the technology, and that point has been passed.

Works Cited

Barnhurst, Kevin G., and Diana C. Mutz (1997). American Journalism and the Decline of Event-Centered Reporting. *Journal of Communication*, 47 (4), 27–53.

Barnhurst, Kevin G., and John Nerone (2001). *The Form of News: A History*. New York: Guilford Press.

Blondheim, Menahem (1994). *News Over the Wires: The Telegraph and the Flow of Public Information in America, 1844–1897*. Cambridge, Mass.: Harvard University Press.

Blondheim, Menahem (2002). "Public Sentiment Is Everything": The Union's Public Communications Strategy and the Bogus Proclamation of 1864. *Journal of American History*, 89, 869–899.

Blondheim, Menahem (2003). Harold Adams Innis and his Bias of Communication. In Elihu Katz et al. (eds.). *Canonic Texts in Media Research*, 156–190. Cambridge: Polity Press.

Carey, James W. (1974). The Problem of Journalism History. *Journalism History*, 1 (1), 1–7.

Carey, James W. (1983). Technology and Ideology: The Case of the Telegraph. *Prospects*, 8 (1), 303–325.

Carey, James W. (2008). *Communication as Culture: Essays on Media and Society*. New York: Routledge.

Carwardine, Richard (2006). Abraham Lincoln and the Fourth Estate: The White House and the Press during the American Civil War. *American Nineteenth Century History*, 7 (1), 1–27.

Castells, Manuel (2009). *Communication Power*. New York: Oxford University Press.

Chalaby, Jean K. (1996). Journalism as an Anglo-American Invention. A Comparison of the Development of French and Anglo-American Journalism, 1830s–1920s. *European Journal of Communication*, 11 (3), 303–326.

Clark, Charles E. (1994). *The Public Prints: The Newspaper in Anglo-American Culture, 1665–1740*. New York: Oxford University Press.

Cowan, Brian W. (2004). Mr. Spectator and the Coffeehouse Public Sphere. *Eighteenth-Century Studies*, 37 (3), 345–366.

Craft, Erik D. et al. (1999). Private Weather Organizations and the Founding of the United States Weather Bureau. *The Journal of Economic History*, 59 (4), 1063–1071.

Czitrom, Daniel J. (1982). *Media and the American Mind: From Morse to McLuhan*. Chapel Hill: University of North Carolina Press.

Digby-Junger, Richard (1996). *The Journalist as Reformer: Henry Demarest Lloyd and Wealth against Commonwealth*. Westport, Conn.: Greenwood Press.

Du Boff, Richard B. (1980). Business Demand and the Development of the Telegraph in the United States, 1844–1860. *Business History Review*, 54 (4), 459–479.

Du Boff, Richard B. (1984). The Telegraph in Nineteenth-century America: Technology and Monopoly. *Comparative Studies in Society and History*, 26 (4), 571–586.

Gitlin, Todd (1980). *The Whole World's Watching*. Berkeley, Los Angeles: University of California Press.

Grier, David A. (2013). Text, Translation, and the End of the United Press. In John Nerone (ed.). *Media History and the Foundations of Media Studies*, 500–520. Malden, Mass.: Wiley Blackwell.

Haller, Mark H. (1979). The Changing Structure of American Gambling in the Twentieth Century. *Journal of Social Issues*, 35 (3), 87–114.

Harvey, David (1989). *The Condition of Postmodernity*. Oxford: Blackwell.

Hay, James, and Jeremy Packer (2004). Auto-mobility, the Transported Self and Technologies of Freedom. In Nick Couldry and Anna McCarthy (eds.). *Mediaspace: Place, Scale and Culture in a Media Age*, 209–232. London: Routledge.

Hochfelder, David (2006). "Where the Common People Could Speculate": The Ticker, Bucket Shops, and the Origins of Popular Participation in Financial Markets, 1880–1920. *The Journal of American History*, 93 (2), 335–358.

Hofstadter, Richard (1955). *The Age of Reform: From Bryan to FDR*. New York: Knopf.

Hughes, Patrick (1980). American Weather Services. *Weatherwise*, 33 (3), 100–111.

Jackaway, Gwenyth L. (1995). *Media at War: Radio's Challenge to the Newspapers, 1924–1939*. Westport, Conn.: Greenwood Press.

Jefferson, Thomas (1984). Letter to John Norvell. In Thomas Jefferson. *Writings*, 1176–1179. New York: Literary Classics of the United States, Inc.

John, Richard R. (2010). *Network Nation: Inventing American Telecommunications*. Cambridge, Mass.: Harvard University Press.

Leonard, Thomas C. (1995). *News for All: America's Coming-of-Age with the Press*. New York: Oxford University Press.

Lippmann, Walter (1922). *Public Opinion*. New York: Macmillan.

Manning, William, Michael Merrill, and Sean P. Wilentz (1993). *The Key of Liberty: The Life and Democratic Writings of William Manning, "A Laborer", 1747–1814*. Cambridge, Mass.: Harvard University Press.

Marx, Karl (1993 [1857/58]). *Grundrisse: Foundations of the Critique of Political Economy*. London: Penguin.

Mattelart, Armand (1996). *Invention of Communication*. Minneapolis: University of Minnesota Press.

McChesney, Robert W. (1991). Press-Radio Relations and the Emergence of Network, Commercial Broadcasting in the United States, 1930–1935. *Historical Journal of Film, Radio and Television*, 11 (1), 41–57.

McKerns, Joseph P. (1977). The Limits of Progressive Journalism History. *Journalism History*, 4 (3), 88–92.

Mindich, David T. Z. (1998). *Just the Facts: How "Objectivity" Came to Define American Journalism*. New York: NYU Press.

Nadler, Anthony (2012). *Making News Popular: Contemporary Models of Popularizing the News*. Ph.D Dissertation: University of Minnesota.

Nerone, John (1994). *Violence Against the Press: Policing the Public Sphere in U.S. History*. New York: Oxford University Press.

Nerone, John (2006). The Public and the Party Period. In Jeremy Packer and Craig Robertson (eds.). *Thinking with James Carey*, 157–176. New York: Peter Lang.

Nerone, John (2008). Newswork, Technology, and Cultural Form, 1837–1920. In Barbie Zelizer (ed.). *Explorations in Communication and History*, 136–156. New York: Routledge.

Nerone, John, and Kevin G. Barnhurst (2012). Stead in America. In Roger Luckhurst et al. (eds.). *William T. Stead: Newspaper Revolutionary*, 98–114. London: British Library.

Preda, Alex (2009). *Framing Finance: The Boundaries of Markets and Modern Capitalism.* Chicago: University of Chicago Press.

Pred, Allen R. (1973). *Urban Growth and the Circulation of Information: The United States System of Cities, 1790–1840.* Cambridge, Mass.: Harvard University Press.

Rantanen, Terhi (2006). Foreign Dependence and Domestic Monopoly: The European News Cartel and U.S. Associated Presses, 1861–1932. *Media History*, 12 (1), 19–35.

Schivelbusch, Wolfgang (1986). *The Railway Journey: The Industrialization of Time and Space in the 19th Century.* Berkeley: University of California Press.

Schudson, Michael (1986). When? Deadlines, Datelines, and History. In Robert K. Manoff and Michael Schudson (eds.). *Reading the News*, 79–108. New York: Pantheon.

Schudson, Michael (1992). *Watergate in American Memory: How We Remember, Forget, and Reconstruct the Past.* New York: Basic Books.

Standage, Tom (1998). *The Victorian Internet: The Remarkable Story of the Telegraph and the Nineteenth Century's On-Line Pioneers.* London: Weidenfeld & Nicolson.

Starr, Louis M. (1954). *Bohemian Brigade: Civil War Newsmen in Action.* New York: Alfred Knopf.

Wheeler, Tom (2008). *Mr. Lincoln's T-Mails: How Abraham Lincoln Used the Telegraph to Win the Civil War.* New York: Harper Business.

Williams, Raymond (1972). *Television: Technology and Cultural Form.* London: Fontana.

Yates, JoAnne (1986). The Telegraph's Effect on Nineteenth-Century Markets and Firms. *Business and Economic History*, 15, 149–163.

Zaloom, Caitlin (2006). *Out of the Pits: Traders and Technology from Chicago to London.* Chicago: University of Chicago Press.

From Handmade to Technology-Driven Journalism: Changes and Developments in Research, Writing, Editing, and Makeup

Jürgen Wilke

Introduction

In communication science, the term "media" usually refers to technical means that (can) serve to distribute information to large numbers of people. Media history has therefore always had a technical dimension as well. This has been valid ever since the invention of the printing press by Johannes Gutenberg in the mid-fifteenth century, which made the mass reproduction and distribution of written works possible in the first place.

This technique gave rise to modern media in the true sense of the word, that is, newspapers and magazines. However, it required not only a means of reproduction but also the creation of the content to be reproduced. For this purpose, individuals first known as newspapermen and later journalists had to take action. They had to perform specific activities for which they required cognitive and practical skills. In journalism, manual operations were sufficient for a long time before technical means were introduced. Initially, this occurred exclusively in the form of written journalism. Centuries later, other techniques were added that would lead to similar (but also other and additional) results. In the following, the long-term technological changes in journalism will be investigated, including the determinants that shaped this process.

(Technical) Levels of Action in Journalism

It is possible to distinguish four levels in journalism on which technological achievements are relevant and have influence over the creation of products (Dooley 2007, 19):

1. The acquisition (research) and transportation of information/news,
2. the processing and treatment of information/news,
3. the content and presentation of information/news, and
4. the distribution of information/news.

These four levels of action will be considered separately in this context, which is justified since they are chronological. However, they do intertwine and overlap. For historical reasons, I will treat these areas or phases primarily based on the example of classic (press) journalism and will then move on to briefly elaborate on the changes brought on later by the new media.

(1) News acquisition

Determination and physical transportation of news

While the *distribution* of news has been linked to technology since the very beginning of the press, its *acquisition* was long accomplished without the use of technology, at least when it is considered to involve the employment of certain devices. When viewing it in a broader sense as a general term for human actions and procedures, then journalism has always been based on technology (for the terminology, Ropohl 2006). However, this rather general sense of the term should be employed here to speak about technology. Two procedures can be distinguished in the first phase or level of the journalistic work process: the determination of information (research) and its (physical) transportation. How news found its way into the early newspapers is an area we know little about. The hard evidence is simply insufficient, unlike that which exists for the emergence and function of printing shops based on woodcuts and engravings, which have been around since the sixteenth century.

News was usually obtained from correspondents present at the scene of newsworthy events or where news arrived from elsewhere and was relayed. How these correspondents exactly went about obtaining this news remains unknown. Possible correspondents included diplomats, salesmen, and academics (Wilke 2010). They must have been provided with oral reports by informants and profited from personal and professional contacts. They presumably obtained some of their news from written sources or at public places (churches, markets). The rialto in Venice is known to have served such a purpose (De Vivo 2007, 89–97). These correspondents may have possibly had access to letters or writings by other authors. Active "research" was probably not (yet) conducted at the time.

The collected information was transmitted to the printers by means of letters. This required physically transporting it over long distances. This was performed by mail institutions, which emerged in the early modern period, in the case of the German Reich predominantly through the Thurn und Taxis postal system (Behringer 2003). Couriers on horseback carried the letters and delivered them to the local postmasters, who then passed them on to the printing shops. The technical facilities of the mail institutions were thus essential to the acquisition of information (and later to the distribution of printed newspapers). The transportation speed depended on the capacities of human and animal strength. A radical acceleration could not be achieved by these means. In exceptional cases, carrier pigeons were used, as in pretelegraphic communication. It was not until the advent of steam power in the early nineteenth century that the transportation speed could be increased over land routes (railway) as well as sea routes (steamboats). Their use resulted in a considerable reduction of transportation times (Kaukianen 2001).

Another technology—pneumatic mail—was created for the same purpose. A network of tubes was installed through which written material of all kinds was propelled in canisters or leather pouches with the aid of pneumatic pressure (Arnold 2000). They were not only used in large German cities (Berlin, Bremen, Cologne, Düsseldorf, Frankfurt am Main, Hamburg, Leipzig, Munich), but also served as in-house delivery systems on company sites (Arnold 2000, 33). As photographs indicate, the Ullstein publishing house had such a system. It connected the various departments and was presumably used for transferring information between the editorial department and the case room (ibid., 62). The current possibilities of transmitting manuscripts in a fraction of a second from one computer to another are in some ways an acceleration of pneumatic mail.

Telegraphy

Decisive technological developments in news acquisition were not made until the nineteenth century. When they finally were, however, it occurred in several ways and in ever-shorter intervals of implementation.

After the discovery of electricity in the eighteenth century, several attempts were made to use it in order to transmit signals over spatial distances. This had already been tried using mechanical devices, such as the optical telegraph (Beyrer and Mathis 1995). However, its capabilities were very limited and it did not gain any real significance for the journalistic acquisition of information. Various kinds of electronic telegraphs were constructed in the early nineteenth century that used very different methods to encrypt and decrypt the transmissions. This problem was not sufficiently solved until the American Samuel F. B. Morse applied for a patent on his device in 1840. A few years later, David Edward Hughes invented the type-printing telegraph, which could print the transmitted text on strips of paper in plain writing. More technical improvements would follow among which switching from electrical wires to radio waves was one of the most significant after the physicist Heinrich Hertz had discovered electromagnetic waves during experiments in 1886. Ten years later, in 1896, Guglielmo Marconi introduced wireless telegraphy.

Telegraphy was used as a technical medium of information acquisition (and distribution) beginning in the mid-nineteenth century, meaning that the transport of information now became nonmaterial. It is a well-known fact that the press quickly drew on this technology and to a growing degree. The telegraph bureaus that emerged in the mid-nineteenth century and would later be called news agencies were used for this purpose. Telegrams became an essential part of a newspaper's content. While one advantage was the speed at which they could be transmitted, they initially involved considerable costs and therefore contributed only a small share of content to newspapers (Wilke 2004a). At first, it was primarily stock exchange reports and price quotations that were transmitted with the aid of telegrams. Such data required no further journalistic treatment. Other kinds of important messages were only gradually transmitted this way. Since the so-called telegram style with its numerous abbreviations was used, no doubt for economic reasons, journalists had to create legible texts from the telegrams. This of course required considerable effort.

In Germany, as in many other countries, telegraphy was organized by the state-run postal service. The connections ran via the post offices and were operated by postal employees. The German hub was the '*Haupttelegraphenamt*' in Berlin, which employed numerous telegraphers. From there telegrams could be transmitted citywide via the pneumatic mail system installed in the last third of the nineteenth century (Arnold 2000). This proved to be faster and less expensive than using carriers. The long-distance pneumatic mail system in Berlin connected the Haupttelegraphenamt with seventy-four pneumatic post offices, local post offices, but also with other institutions (the stock exchange, the Reichstag, general staff) and, for example, '*Wolffs Telegraphisches Bureau.*' Courier services were still needed to deliver mail to the final recipients.

The obstacles that telegraphy still faced in its beginnings were overcome in the course of the twentieth century with the development of the teleprinter. It combined a keyboard with a printing mechanism and was thus able to deliver plain text. It was connected with telephone lines via electric wires or radio connections, if necessary on leased lines. Such devices could then be installed in the newsrooms, which effectually eliminated spatial distances in news transmission.

The telephone

After the invention of telegraphy, it did not take long before another technology for acquiring journalistic information was developed: the telephone. The American Alexander Graham Bell first devised a version ready for production in 1877. In 1900 there were 290,000 telephone connections in Germany; ten years later this number had already more than tripled (Rammert 1990). Only in the United States did the numbers grow at a faster rate. It is no surprise that such devices first took hold in industrial companies. Photographs indicate that such devices could also be found in the newsrooms of large newspapers (Fig. 1). The extent to which they were used for acquiring information at the time is unknown. In many cases, they were in-house telephones that replaced the internal communication that often required more staff. Initially, telephones could only be used to place local calls, while at the same time necessitating a switchboard and local staff in charge of dispatching. In the eighties women still predominantly performed this job. Using the telephone was somewhat inconvenient as long as the instruments were still wall-mounted. Talking on the phone became much more comfortable with the advent of desktop telephones, which could be used while sitting.

In the field of journalism, the telephone could be used for a variety of purposes, in particular for direct research. More importantly, in the beginning it was probably a mere transmission resource in "real time." A correspondent or journalist in the field could transmit his report orally to the editorial staff, which either recorded it or perhaps even turned it directly into copy. Unlike the telegraph, the telephone was less expensive and easier to use. As indispensable as the telephone would become to journalists, and as much as it simplified field research, the technology was still confined to specific

Fig. 1: Hermann Dupont, editor-in-chief of the Berliner Morgenpost, with a desktop phone (c. 1900) (Source: 50 Jahre Ullstein, 161)

places. This continued to be the case as long as journalists were compelled to use landlines. It was not until the invention of the mobile telephone in the second half of the twentieth century that journalists became more flexible. The first mobile phones in the sixties were still very unwieldy. The A-net telephone default device B 72 weighed 16 kilograms and came at a retail price of what is today equivalent to between 4,000 and 7,500 Euros. The mobile phone's actual breakthrough (at least in Germany) first came with the introduction of the fully digitalized D- and E-nets (in regular use in 1993/94). It was not until the instruments were miniaturized that they could be used on a widespread basis for mobile communication, enabling true mobility (Wilke 2004b, 24–27).

Research sources

There were, and still are, other technologically advanced means of acquiring information than those described above. One of them has always been the analysis and evaluation of newspapers. This practice was witnessed quite early and was also criticized. It was a logical choice to the extent that, and as long as, (original) sources were in short supply and time pressure was low (for example, due to the longer intervals between publications). In addition, news was not yet protected by copyright.

Just how natural the use of other newspapers for one's publications really was becomes evident in literature related to practical journalism. As Julius David writes in his *Berufsführer* (1902), a career guide, the editorial rooms received "piles of correspondence and newspapers" each morning. They were skimmed through, and "articles or announcements that somehow seem to be in need of editing are circled in red or simply cut out" (David 1902, 109). A red marker and scissors thus become "technical" instruments in addition to the glue for pasting the clippings together. This practice, used not only by journalists but also by those who read newspapers, led to the phrase "cut and paste," which is still in use today (te Heesen 2002). It was practiced by smaller newspapers with a limited number of editorial staff members. In the *Journalistischer Führer (Journalistic guide)* from 1899, the anonymous author writes the following: "Most people believe that the instruments with which an editor should have to work are ink, quill, and paper. In reality however, scissors and glue play a much greater role in an editor's life …" (ibid., 34).

A great deal has changed in the meantime. However, even though contemporary journalists no longer reprocess other newspapers or press organs, as their forerunners did, they still use them to inform themselves and to research and select topics. In Germany, for this purpose reference is made primarily to prestigious and high-quality newspapers (Reinemann 2003). For a long time, many newsrooms had television sets in order to follow the news and live reports. Today, copying information from other media sources is regarded as being against the rules if it is done without naming the sources. Yet increased computerization has made it easy to simply mark digital source content with the help of a mouse, then cut, copy, and paste it.

Other means of acquiring news were still available in editorial rooms. As shown in blueprints of newspaper publishing houses in the late nineteenth century (50 Jahre Ullstein 1927; Freyburg and Wallenberg 1977), one room was reserved exclusively for the editorial library, "the most important tool of a journalist" (Wehle 1883, 43). Journalists had access to dictionaries, reference works, atlases, historical works, legal literature, and so on in order to verify information. Individual editors could also access smaller personal libraries at their workplace.

The installation of newspaper archives has achieved even greater relevance as a research tool since the nineteenth century (Englert 1984; Schulz 1987). While newspapers in archived form were long regarded as rather unimportant, a change in attitude took place when it became common to systematically and continuously cut, catalogue, and store relevant articles from one's own or other publications. Scissors and glue were once again the tools required for this task. The first newspaper archives emerged in the nineteenth century on personal initiative. Larger newspapers, such as the *Neue Züri-*

cher Zeitung, the *Hamburger Fremdenblatt*, or the *Frankfurter Zeitung*, began creating such archives. The most intense form of archiving was done by the Ullstein publishing house in Berlin (50 Jahre Ullstein 1927; Freyburg and Wallenberg). Photographs show a large room filled with a circular arrangement of index card cabinets in which the archived material was arranged according to a (non-identifiable) system (alphabetically and/or according to subject) (Fig. 2). However, small or medium-sized publishing houses could hardly afford anything of this kind.

Fig. 2: Text and image archive at the Ullstein publishing house (Berlin, c. 1900) (Source: Freyburg and Wallenberg 1977, 175)

The examples cited above as well as the experience gained in the United States and the United Kingdom sixty to eighty years earlier had an impact on the creation of newspaper archives in Germany after 1945. Larger publishing companies were again at the forefront of this development. The problem associated with archiving was intensified by the large amount of material at hand and the simultaneous need for storage space. The workload increased, which in turn led to differentiating between material with a documentary or journalistic purpose, wherever this was possible. At the same time, the role of the archivist changed from performing a service to controlling information (Schulz 1987, 66). In order to meet the challenges of the archival and documentary tasks, use was made of the range means of storage, administration, and development available since the sixties, from the microfilm to modern forms of digitalization. Not all of these solutions proved to be sustainable, such as, for example, "electronic scissors," which failed due to the changeover to punch-card technology in the media in the early sixties (Schütz 1963).

Research, of course, was not performed exclusively in one's own editorial archive. As early as in the nineteenth century, private individuals had created collections of clippings as a source of information. The first professional services were established in Germany after the turn of the twentieth century. Economic institutes set up archives whose use was made possible with the aid of mail services or the telephone. An effective change in this did not occur until the sixties. For one thing, the new electronic data processing and storage possibilities were developed into comprehensive information systems (for example: GBI-Genios German Economic Database). At the same time, the use of computers, through which a direct connection to these information systems was possible, came to be more widespread. With the advent of the Internet, journalistic research became even more extensive and unrestricted.

(2) The processing and treatment of news

Manual writing devices

What is true for the acquisition of news is even more so for its processing and treatment: These phases of the journalistic production process were for the most part devoid of technology for long time. What was to be printed in a newspaper had to be handwritten before it could be typeset and then duplicated. For many centuries, the main writing instrument was in fact the quill. In an engraving from the eighteenth century, a newspaperman can be seen holding a quill (Fig. 3). For centuries, there was no technological development in this area except for slight modifications to the quill. While in the Occident most people preferred using quills made from goose feathers, the eighteen-thirties and -forties saw the production of quills with a steel nib (Vaupel 1987), which became more prevalent as they became flexible enough to no longer cause damage to the paper. Journalism was bound to this archetypical style of text production for a long time. For informal use (taking notes, stenography), a lead pencil was used as an editor does one foto in from late nineteenth century (Fig. 4).

Fig. 3: A "newsroom" in the early eighteenth century: A post rider delivers a letter to a newspaperman, identifiable by the quill (Etching from 1701)

Fig. 4: Ernst Wallenberg, editor-in-chief of the B.Z. am Mittag (Berlin), writing with a pencil (c. 1910) (Source: 50 Jahre Ullstein, 197)

The quill also required the writer to maintain a relatively stationary position, since ink was needed. The acidic ink that was initially used and the fact that the quill had to be dipped into the inkwell after every few letters or words turned writing into a rather uncomfortable activity. This circumstance was also alleviated with the aid of a technological improvement: the fountain pen with an integrated ink sac. While a prototype had been produced as early as in the seventeenth century, it was not until the nineteenth century that its use became widespread with the addition of components such as gold nibs and a feed, which ensured a steady and uniform ink flow. The fountain pen increased writing capabilities in manifold ways. This increase became even more apparent with the advent of the ballpoint pen, which transmits ink from a reservoir inside the shaft onto paper with the aid of a rolling ball inside the tip of the pen. Patents on this technology date back several centuries. However, mass production of the ballpoint pen did not begin until the nineteen-thirties and -forties. While in 1950 the first models retailed at a price of what is today equivalent to ten Euros, prices soon fell sharply, prompting the ballpoint pen to become a universal and flexible manual writing instrument.

The typewriter

The mechanization of manual text production was achieved through the construction of the mechanical typewriter, the result of numerous preceding inventions, the earliest of which can be traced back to the early eighteenth century. The first typewriter was industrially produced in Copenhagen in 1867 (Müller 1900; Dingwerth 2008; Dotzler and Hitzenberger 2009). Several types of typewriters were manufacturing beginning in the eighteen-eighties, all of which had different mechanisms to tackle the problem of the letters being transferred to paper in a random order. Keyboard typewriters and index typewriters were two basic types that could still be subdivided into a number of varieties, and all of them boasted several different technological improvements. One major development was the frontstroke mechanism invented by Franz X. Wagner, which unlike the understroke machines allowed what was being typed to be visible to the typist.

The typewriter was predestined for processing journalistic information. It not only accelerated the speed of writing, in particular after its adaptation to the ten-finger method; a further advantage was that typesetters, who had previously had to deal with numerous individual styles of handwriting, now benefited from a uniformly legible font. In any case, the use of mechanical typewriters became quite widespread in publishing houses by the late nineteenth century, spearheaded by the United States, where most of the typewriters were produced.

It is remarkable that typewriters are rarely seen on any of the editors' desks in photographs of editorial rooms in Germany at the turn of the twentieth century. However, the contrary seems to be true in the commercial departments of publishing houses (Freyburg and Wallenberg 1977, 4, 49). Does this perhaps indicate that typewriters were not used by journalists back then? This is probably the case, since the introduction of the typewriter coincided with the creation of the secretarial profession. Thus this device can to a certain extent be viewed as a "tool of emancipation" (Schiffhauer, V 8). Typing remained women's work in newsrooms and was also one of their duties in editorial offices; they took down dictation in person or via the telephone (on headphones), except when they had been handed handwritten notes. There is photographic evidence of this, at least in Germany, dating from around 1930 (Fig. 5). Some of the female writing staff even performed journalistic tasks. In his *Praktische Winke für Schriftsteller und Korrespondenten (Practical handbook for writers and correspondents)* from 1905, Heinrich Kreiter recom-

Fig. 5: Workplace at the Berliner Börsenzeitung with a female secretary sitting at a typewriter (c. 1930) (Source: Bertkau 1930)

mended "getting a typewriter or [...] having manuscripts copied on one of them" (Kreiter 1905, 13).

In the twentieth century, the typewriter evolved in particular in two ways. The first typewriters were rather large and heavy—they could weigh up to fourteen kilograms, which also limited the writer in her/his mobility. This called for the production of smaller, lighter, more mobile typewriters, if possible also faster and generating less noise. This was achieved in part in the first decade of the twentieth century, when several models were developed to this end, among them the portable typewriter, which usually came in a small case that journalists could carry with them when traveling.

As is the case for all technological developments, improvements on the typewriter aimed at increasing the speed of typewriting while at the same time allowing for more comfort. Both of these were accomplished with the change from the mechanical to the electric typewriter. A patent for the first such device was applied for as early as in 1913, yet market maturity was not achieved until they were fitted with exchangeable typeballs in the early sixties. While this typewriter model achieved great popularity, it was likewise soon superseded.

The computer

The reason for the decline of the typewriter was the development of the computer as a universal tool for text input and text processing. Computer-based data processing also developed via several substages and was accelerated by the invention of the microprocessor. From the seventies onward, computers were smaller and easier to use, and they found their way into all areas of industry, business, and private life. Software upgrades facilitated a large number of technical operations, not only in text processing but also in additional methods that far surpassed the old typewriter. Even the hardware could then be minimized, leading to the ubiquitous laptops, notebooks, and BlackBerry devices in use today.

The most interesting question is how technological development from the quill to the word processor has had an impact on the content and form of journalistic production. Research on the psychology of writing indicates, for example, that the latter has increased the flow of writing without having to fear that copy cannot be revised. The speed of writing corresponds with the speed of thinking. Furthermore, a lot less is lost while writing, as awareness of the content remains intact through the visibility of the text. Word processors also mean that the writer has more time to change or revise a text and thus increase the quality of the final copy (Kellogg 1994, 142–160).

(3) News content and formal presentation

Technology has also influenced the content and presentation of news over the course of journalism history. The flow of news had been more of a "runlet" for a long time, because acquiring and transmitting it was cumbersome and time-consuming. It was not until postal services became more widespread that this runlet grew, which also led to a wider range of themes in the news. A newspaper's capacity continued to be limited by the Gutenberg printing press, which only allowed the production of a limited number of copies (especially in light of the pressure associated with having to be up to date). Remarkably enough, this did not change for centuries. It was not until the invention of the rapid press in the early nineteenth century that the newspaper was relieved of its shackles. It not only increased the production speed of newspapers and allowed for a larger number of copies to be printed, it also facilitated an increase in newspaper formats. However, the production of the rapid press progressed rather slowly, and its use was for the most part limited to large publishing houses, including in Germany.

Technological advances since the nineteenth century have not only increased the amount of information, thus adding a great deal of new content to the newspaper, they have also had an influence on the journalistic form. Telegraphy required that its content be very brief, giving rise to the newsflash format, which was initially considered a category of its own. In the United States, the "inverted pyramid" news format is attributed to this technology, a claim that has meanwhile come under scrutiny (Pöttker 2005).

In the early days of the newspaper, the typesetter had to make do with simply aligning the columns of news and furnishing them with headlines that indicated the origin of the transmission and when it was received. This form of presentation did not change for centuries. As late as the nineteenth century, the content was merely placed in categories with headlines of their own, e.g. *'Parlamentssachen' (parliamentary material)*. It was not until much later that captions and headings were added to newspaper articles, and even then only sporadically until the early twentieth century. At about the same time, illustrations were introduced (which had already had been the case with magazines for decades, however). This was also due to technical reasons. Illustrations in newspaper print did not become possible until the invention of the halftone print (Georg Meisenbach 1882; Peters 2007). Illustrated magazines were produced with the help of wood engravings. More innovation was still required in the field of photography, the instantaneous photograph for example.

(4) The makeup and duplication of news

As already pointed out, the distribution of journalistic news was characterized and facilitated by the use of technology. For centuries, this was done by using print, letters shaped from molten lead. Until far into the nineteenth century, newspaper pages had to be typeset by hand, line for line, for which professional typesetters were responsible (some of whom would eventually cross over into journalism). Their job consisted in "materializing" the handwritten manuscripts. Despite the fact that they developed remarkable skill in the process, it continued to be time-consuming and complicated. The efforts made to find a solution to this problem led to the invention of the Linotype in the eighteen-eighties by Ottmar Mergenthaler (1854–1899) (Schröder 1941). It enabled typesetting an entire line with the help of a keyboard, and thus greatly reducing the time it took to create entire pages. The first machine, called the "Blower," was followed by numerous other models. Typesetters were now in charge of operating such machines. In the case of the Monotype, introduced by Tolbert Langston in 1897, the molding and setting processes were again separated. The typesetter created a punched card with the aid of a keyboard, which in turn operated the typesetting portion in the form of single letters, which came together into lines. Headlines and captions had to be set by hand until well into the twentieth century.

The procedures and circumstances described above meant that journalists had to carry out part of their profession on the threshold to technical production in the typesetting room (Fig. 6). They handed over their copy to the typesetters, who would then create the metal type. The pages of copy could only contain writing on one side, as they were cut into several pieces (especially if they were somewhat long) and rearranged by the typesetter. The rearranging and formation of the individual pages, called page makeup, was completed in the composing room by makeup men. Whether or not a sufficient amount of type—or perhaps even too much—was created in the course of producing a newspaper was tested by the makeup man with the help of a string, which would later be called *'Spiegel' (page plan or dummy)* (Wehle 1883, 11–15). While the early newspapers were still made up in left- and right-justified format, some newspapers in the eighteenth century already had a two-column page makeup. Page makeups of three to five columns were introduced with the enlargement of newspaper formats in the nineteenth century.

Fig. 6: Makeup of newspaper pages by means of metal types (c. 1960) (Source: Journalist 9/1977, 16)

The editor-in-chief usually supervised the page makeup. As far as it had not already been arranged with the makeup man, the appearance of the newspaper pages was designed. The clichés of the pictures were to be fit in. While the makeup man arranged the metal sorts and moved them back and forth if necessary, the editor-in-chief, who was familiar with the articles' content, decided on their correct placement. But the makeup man also had to keep abreast and adhere to certain (typographic) rules for the page makeup. He was not allowed to exchange or mix up columns and articles (Wehle 1883, ibid.). The arrangement of the newspaper pages was of course standardized to a certain extent (especially the front page). After the page makeup was completed, the printing plate was locked in a frame so that it could be moved and put into the printing press, operated by the printer. With the invention of rotary printing, another production phase was added, so-called reprography. Its function was to transfer the plain printing plates onto (semi-) round metal or plastic plates, which in turn could be mounted onto the rollers of the printing press.

In the seventies and eighties, drastic revolutions took place in the field of printing technology (Ratzke 1982). Photo- and filmsetting almost completely replaced lead typesetting, the latter surviving only in some niches. Photo- and filmsetting was developed in the fifties in different systems: either by the screening through of documents (stencils) or the creation of digitalized, recorded templates. In both cases, cathode ray tubes were used as light emitting sources. This typesetting procedure also changed the page makeup. At first, the new procedure remained almost analogous to lead typesetting, only now screens were glued together and "mounted" on moveable walls or tables and then used to create printing plates with the help of exposure to light (Fig. 7). Compact systems quickly followed and consisted of a keyboard and screen for text composition, with a small window or screen for visual control as well as storage and output units. The editorial systems introduced in Germany during the eighties were a direct result of this configuration. The electronic page makeup for entire pages, which was completed with the help of devices with a screen, rendered the complicated gluing together of completed pages unnecessary (desktop publishing).

Fig. 7: Makeup of newspaper pages by means of phototypesetting (1980s) (Source: Journalist 1/1978, 21)

The development described above had tremendous consequences for journalism (Weischenberg 1978; Arbeitsgemeinschaft für Kommunikationsforschung 1983; Mast 1983). An end was put to the old separation of journalistic and technical features (typesetter). This was accompanied by in part aggressive confrontations over journalistic responsibilities and tasks in Germany. Journalists, it was argued, would be "degraded" to technicians, and the profession of the typesetter rendered irrelevant. The traditional journalistic self-image in Germany was thus being challenged. Labor unions in particular voiced their opposition (Prott 1983). Only after long confrontations could the disputed questions be settled through a new collective wage agreement. Today, when the mechanization of the journalistic profession has progressed even further, the difficult birth of the electronic newsroom is difficult to comprehend.

Changes Associated with the New Media

Radio

The relevance of technology for journalism and the changes caused by it have been described above primarily based on examples from the realm of the press, that is, the print media, in particular the newspaper. New kinds of media emerged in the twentieth century, at first in the twenties in the form of radio broadcasting. Thus with the exception of film, a new technology for the mass transmission of information had become available for the first time in centuries. Radio technology changed the *distribution* of information via radio waves. Messages were no longer needed in the form of a digital system of symbols on a hard carrier and thus transported mechanically, but could now be transferred as audio signals, that is, sound waves. At the same time, technical methods became more complex and required elaborate facilities.

In radio broadcasting, a basic distinction must be made between recording and broadcasting technology (Goebel 1950). The former (like the printing press in print media) is a matter of a specific broadcasting staff that is responsible for the broadcasts. This can be said in the same measure for recording technology, but this also directly influenced journalism as such. However, there were two reasons why the impact on the latter was initially limited. For one thing, journalists who dealt with current affairs only played a secondary role in early German radio broadcasts (Heitger 2003). Radio broadcasting was mainly installed as an educational and entertainment medium and dominated by spoken-word and music programs. The so-called *'Drahtlose Dienst AG' (DRADAG) (wireless service)* was in charge of the news for all German broadcasting facilities. At first, radio newscasts were not too different from news in the print media, the only difference being that there were verbal presentations of the news. Initially, only agency and press releases were read. A parity press review was introduced in 1928 (Stohl 1931, 123–124). But where individual newscasts or other content for radio broadcasting were produced, written texts were also being created that would then be read openly, as was the case for specialized spoken presentations such as farming and religious programs or programs for women, and so on. In the course of this, rules of conduct for radio broadcasts were introduced, keeping in mind that hearing is very different from reading. "Audibility" was tested with the help of a special device: "When the first draft of a newscast has been typewritten, it is dictated into a parlograph and then listened to by the author, or,

Fig. 8: Voice booth in a radio station (Mitteldeutscher Rundfunk, Dresden, c. 1930) (Source: Lieberwirth 2000, 47)

even better for the evaluation, by another editor. This way, the actual effect of the work can be predicted in advance" (Stohl 1931, 128).

Recording studios were needed, as were rooms for the editors in charge of providing the programs. For moderated programs and presentations, a voice booth with a microphone was sufficient (Fig. 8). The microphone was the main piece of equipment in audio recording, different types of which were developed and tested in the twenties/thirties (Goebel 1950, 389–392; Lieberwirth 2000, 135–142). Noise insulation and noise levels in particular were of interest in this process, but also sound quality. Microphone amplifiers and level meters were needed for this purpose. In addition, microphones were rather unwieldy at first and were therefore reserved for studio use only. Smaller and more compact devices enabled transmitting from outside or delivering live reports, journalistic innovations, so to speak, inherent in the medium. Yet the groundwork of broadcasting journalism remained primarily based in writing, and to a certain extent the typewriter continued to be an important tool for radio journalists (Maibohm 1948).

Because the radio was not a storage medium, radio programs were broadcast live directly via the microphone. Further technological advances were required in order to store and edit audio recordings. One possible solution (apart from musical recordings on shellac and wax photograph records) was discovered in the thirties in electromagnetic form: the magnetic audiotape (Lieberwirth 2000, 183–194). Their playback devices were replaced in later decades with less bulky cassette players, thus enabling the ongoing use of live quotes.

Television

Radio was quickly followed by another medium: television. The transmission of moving images was based on a rather complex technological framework and passed through several stages, with countless steps and patents, from mechanical to electronic image resolution (Goebel 1953; Abramson 1987). More inventors and engineers were involved in this process than had been in the development of previous communication technologies. In the field of television, a distinction also has to be made between recording and broadcasting technology. While television broadcasting technology in a certain way corresponds to that of radio broadcasting, the recording technology had become much more complicated in comparison to the older medium. It included cameras, devices for image leveling and direction, as well as image editing, all of which are production stages that lead to the creation of new professions, such as cameramen, editors, and cutters. The technology- and staff-related expenses, which had been relatively reasonable in radio broadcasting, increased manifold and presented journalists with new challenges.

In television, older media also shaped journalistic tasks. News had to be read by a—now visible—speaker. The presentation of moving images oriented itself toward the live reporting techniques of the weekly newsreels produced for the cinema. The intermediate film system, which had been around since the thirties, was used in this case. Cameras recorded events on filmstrips, the electrochemical image carrier that gave rise to a new mass medium in the late nineteenth century. These strips of film were then recorded by a television camera and broadcast on television screens. These films had to be transported mechanically during events that took place in the studio, which was still the case for the *Tagesschau,* the news program in the early days of postwar German television. Newscasts by foreign correspondents had to be flown to Hamburg, which meant that there was a considerable time lag in reports from abroad. This did not change until the advent of the transmission of electric signals via wires (cables) and finally via radio waves (satellites).

Several basic forms of journalistic work processes remained in use in television journalism, such as, for example, telephonic research, the use of agency news material, or writing and editing texts on paper. But beginning in the fifties, numerous innovations in the area of production technology had a major impact on the journalistic product (Heim 1991). These included the electronic camera with exchangeable lenses; the mobile and portable camera, which initially was only capable of producing black-and-white images; as well as devices such as the teleprompter. The magnetic video tape recorder (VTR) enabled the storage and multiple use of television images. The *Tagesschau* began using this technique in mid-1959. The detour via filmed material was thus no longer necessary and could be abandoned (Bewilogua and Nieland 1996). The VTR technique has since been replaced by digital recording technology. Electronic broadcasting allowed for the production of live television shows. In fact, electronic coverage made the production of live reports on television possible in the first place. Thus the medium became more mobile, a development that was also furthered by the invention of the "fly-away," which enabled satellite transmission from anywhere (ibid.; Livingston and Van Belle 2005). Numerous technical developments in image direction (vision mixing systems, special effects) have changed the visual presentation of television images and made their composition more complex (Renner 2007). Traditional analog technology has been replaced almost entirely by digital technology.

The Internet

The most recent innovation in media technology, the Internet, shall only be discussed briefly here. It has, however, drastically transformed journalism, and continues to do so (Meier 2002). It has evolved from being exclusively a military application to being accessible by virtually anyone in the world (Abbate 1999). Its spectrum of possible functions has steadily increased over time. Apart from email for the purpose of personal communication, collective forms of use were added, such as the World Wide Web as well as its interactive form, the Web 2.0. The Internet affects all stages of the journalistic production process. It has expanded the possibilities of information acquisition to almost limitless dimensions, while at the same time accelerating them. Computer access facilitates the processing of information, while multimodal forms of presentation have lead to new hybrid formats (Bucher, Lohning and Lehnen 2010). The technical requirements for journalists have increased, yet at the same time new functions have spun off from journalism, for example the Webmaster.

Historical and Systematic Summary

In conclusion, taking a look back at the centuries-long history of the relation between technology and journalism, in the following different historical and systematic aspects will be summed up and grouped under three headings.

Historical development

To begin with, some conclusions can be made with respect to the historical process during which mechanization began to take hold in journalism:

1) Mechanization as a later historical phenomenon

The mechanization of journalism has to be seen as a later historical phenomenon. After centuries of journalistic production without the aid of technical devices, mechanization did not set in until the second half of the nineteenth century. The technological history of journalism did not occur in isolation but is linked with the overall development of technology, society, economy, and culture, which, of course, affect each other as well (Rühl 2000).

2) Acceleration of technology

Just like in other areas of modern technology, the mechanization of journalism has been affected by an accelerated dynamic. Inventions and technological innovations followed each other in ever-shorter intervals, thus accelerating the technological permeation of (journalistic) work processes. However, this dynamic cannot be explained in a teleological way, or from end to beginning, so to speak, as there also have been numerous technological failures and mistakes,[1] for example the picturephone (Lipartito 2003).

3) "Thrust dynamic"

Technological development in journalism (just like in any other field) did not take place in a linear and consistent process but in thrusts and leaps. Techniques that took hold involved the improvement of details or partial optimization. At certain moments, however, innovations revolutionized existing operations and elevated them to a new level. In this case it is justified to speak of turning points in media history, one example of which is the transition from the typewriter to the computer.

Causes and determinants

In terms of history, the mechanization of journalism has various causes and is based on several nontechnical determinants.

1) Physical and physio-chemical determinants

The technical innovations that permeated journalism were decisively affected by physical or physio-chemical discoveries. Simply put, their development occurred in three stages: from mechanical to electric to electronic.

2) Economic determinants

In order for these technological innovations to enter the journalistic process, the physical discoveries had to be made economically applicable in practical form, which was performed by entrepreneurs. Innovations initially called for investments; their use, however, promised economic gain, higher revenue (due to more attractive and successful products), or at least higher savings (due to decreased expenses, for example in terms of personnel). It was primarily economically strong media companies that could afford technical innovations and thus achieve economic gain ("economies of scale"). With a

1 A picturephone is an example for such a failure which repeatedly has been mentioned in media history literature (Lipartito 2003). But even this failure has recently been resolved by the Skype technology in the Internet.

decline in prices as a result of mass production, smaller companies gradually benefited from technological advances.

3) Social and cultural determinants

In many respects, the introduction of technologies in journalism was also determined socially and culturally. Not only in the general sense, namely that being receptive to technology varies in different cultures and societies depending on whether they are traditional or modern. This also applies to journalism. The self-image of German journalists, which is based in a literary tradition, seems to have been an inhibiting factor in the introduction of new technologies, as was the case with the typewriter in the late nineteenth century. This phenomenon repeated itself a quarter of century later, when new editorial techniques became available.

Results and secondary effects

The increasing mechanization in journalism had numerous consequences and was accompanied by far-reaching secondary effects.

1) Increased complexity

The technology used in journalism ultimately became more complex. The inventions that fostered the development and the devices that were needed required ever more subtle procedures and posed increased demands on journalists and users alike.

2) Acceleration

In journalism, new technologies usually led to acceleration, especially with respect to production and distribution processes. The driving force behind this development was the product's inherent requirement of currentness. In order to increase it, distances had to be shortened and overcome, that is, obstacles had to be minimized; not only in terms of transmission, but also by accelerating the processing of information with the aid of new technologies.

3) *Increased capacity*

The use of new technologies in journalism resulted in an increase of capacity. This applies to all phases of the production and distribution process. This of course refers to the acquisition and processing of ever-larger amounts of information. As long as the news flow was only a "runlet," little technological support was needed. With the "information avalanche," however, the use of technological support was unavoidable.

4) *Multimodal presentation*

Journalism remained unimodal for a long time, because news was only transmitted in printed language, that is, digital characters. Early printed newssheets included illustrations (woodcuts), and thus also the mode of iconic characters. However, this practice was discontinued with the transition to intermittency and the inherent expansion of the newspapers' content. Only the technological inventions at the turn of the twentieth century enabled illustrating the medium. Computer hardware and software have expanded the visual spectrum and turned newspapers, and to a much larger degree the new media, into multimodal products.

5) *Blurring of borders*

Another effect of mechanization was the blurring of borders in journalistic production. This is not only true with respect to quantity but to reach. The globalization of journalism could not have been possible without technology.

6) *"Liberation," mobilization*

Several technological achievements contributed to a "liberation" of the journalistic process from a fixed location and increased a journalist's mobility. Perseveratively journalists did work sitting (if not standing in earlier days) at a desk, using incoming material and preparing this material for publication. With new technologies, previously existent constraints were decreased. The electronic media finally enabled live broadcasts.

7) Dematerialization

In several phases of its history, journalism was affected by a dematerialization in the technical sense. This occurred with the advent of telegraphy, as far as the transportation of information is concerned. The same can be said for later transmission technology. The introduction of the computer basically meant the end of paper as an information carrier. Lead type becoming obsolete with the transition to photosetting meant dematerialization in truest sense of the word. Paradoxically enough, this process was also accompanied by a re-materialization, since the immaterial processes also predisposed material devices or machines such as the telegraph, the computer, or the photosetting machine.

8) Increased comfort

One motivation for the headway mechanization made in journalism that cannot be underestimated is increased comfort. Once again, it was about simplifying ever more complex technologies in order to increase the operability of devices, that is, the comfort of their use. This was achieved by a very efficient miniaturization of devices such as the (mobile) telephone, the computer, and so on.

9) The division of labor and (re-)integration

The mechanization of journalism had several adverse effects as well. In the United States, it led to the division of labor between the acquisition of news by reporters and its processing by editors, a phenomenon that had been occurring since the nineteenth century (Salcetti 1995, 52). However, this did not take place in Germany, where a more wholistic idea of journalism prevailed, which is less based on disicion of labour. Yet there was a distinction between journalistic and technical functions. On the other hand, the introduction of editorial technology in the twentieth century caused a (re-) integration of journalistic and technical work processes. This is best demonstrated by the example of the unification of text processing and page makeup on screen devices.

10) Gendering

For centuries, journalism was a profession largely dominated by males. It seems to have been through the use of technology that women could gain a foothold in this profession, albeit initially only in supporting positions. It was not unusual for the widows of printers to continue the operation of their late husbands' shops, thus facilitating the publication of newspapers as early as the seventeenth century. In the late nineteenth century, women were hired as telephone operators in publishing houses. They would soon also work as secretaries and typists. From there, some of them also found their way into active journalism.

11) Increased performance and quality

All in all, the mechanization of journalism led to an increase in performance and quality. The aforementioned effects could already be summed up in this way. But other elements can also be included, such as the quality of research, the quality of information, and so on.

12) Increased level of competence

The demands on journalists increased along with the use of the technologies that were added to the profession in the course of its development. Today, in the age of computers and digital technology, a higher level of competence is required than ever before, especially since manual skills have become less relevant.

Works Cited

Abbate, Janet (1999). *Inventing the Internet.* Cambridge, Mass., London: The MIT Press.

Abramson, Albert (1987). *The History of Television, 1880 to 1941.* Jefferson, N.C., London: McFarland & Co.

Arbeitsgemeinschaft für Kommunikationsforschung (1983). *Schlussbericht: Die Tageszeitung im Prozess der Elektronisierung.* München: AfK.

Arnold, Ingmar (2000). *Luft-Züge: Die Geschichte der Rohrpost in Berlin und anderswo.* Berlin: Gesellschaft für Verkehrspolitik und Eisenbahnwesen (GVE) e.V.

Behringer, Wolfgang (2003). *Im Zeichen des Merkur: Reichspost und Kommunikationsrevolution in der Frühen Neuzeit.* Göttingen: Vandenhoeck & Ruprecht.

Bertkau, Friedrich (1930). 75 Jahre Berliner Börsenzeitung: 1. Juli 1855–1. Juli 1930. In Friedrich Bertkau (ed.). *75 Jahre Berliner Börsenzeitung,* 9–54. Berlin: Berliner Börsenzeitung Druck und Verlag.

Bewilogua, Jürgen, and Jörg-Uwe Nieland (1996). Von der "Arriflex" zum "Fly-away." Zur Technikentwicklung bei der ARD/ARD-Aktuell und beim ZDF/Redaktion Aktuelles. In Peter Ludes (ed.). *Informationskontexte für Massenmedien,* 51–95. Opladen: Westdeutscher Verlag.

Beyrer, Klaus, and Birgit Susann-Mathis (eds.) (1995). *So weit das Auge reicht: Die Geschichte der optischen Telegraphie.* Karlsruhe: G. Braun.

Bucher, Hans Jürgen, Thomas Gloning and Katrin Lehnen (eds.) (2010). *Neue Medien, neue Formate. Ausdifferenzierung und Konvergenz in der Medienkommunikation.* Frankfurt, New York: Campus.

De Vivo, Filippo (2007). *Information and Communication in Venice: Rethinking Early Modern Politics.* Oxford, New York: Oxford University Press.

Dingwerth, Leonhard (2008). *Historische Schreibmaschinen: Sammlerträume; Geschichte, Technik und Faszination.* Regenstauf: Battenberg.

Dooley, Patricia L. (2007). *The Technology of Journalism: Cultural Agents, Cultural Icons.* Evanston: Northwestern University Press.

Dotzler, Bernhard J., and Ludwig Hitzenberger (eds.) (2009). *Schreiben und Rechnen: Eine Technikgeschichte der Informationskultur.* Regensburg: Universitätsverlag Regensburg.

Englert, Marianne (1984). Geschichte und Aufgabenstellung der Pressearchive. In Hans Bohrmann and Marianne Englert (eds.). *Handbuch der Pressearchive,* 7–19. München, New York, London, Paris: K. G. Saur.

Freyburg, W. Joachim, and Hans Wallenberg (eds.) (1977). *Hundert Jahre Ullstein,* 1877–1977 (4 Volumes). Berlin: Ullstein Verlag.

50 Jahre Ullstein. 1877–1927 (1927). Berlin: Ullstein Verlag.

Goebel, Gerhart (1950). Der Deutsche Rundfunk bis zum Inkrafttreten des Kopenhagener Wellenplans. *Archiv für das Post- und Fernmeldewesen,* 2, 353–454.

Goebel, Gerhart (1953). Das Fernsehen in Deutschland bis zum Jahre 1945. *Archiv für das Post- und Fernmeldewesen,* 5, 259–391.

Heitger, Ulrich (2003). *Vom Zeitzeichen zum politischen Führungsmittel: Entwicklungstendenzen und Strukturen der Nachrichtenprogramme des Rundfunks in der Weimarer Republik.* Münster: LIT-Verlag.

Heim, Mark (1991). *Wandlungen der Fernsehproduktionstechnik und ihre Auswirkungen auf die Fernsehproduktion am Beispiel des Südwestfunks.* Master thesis: University of Mainz.

Jacobi, Richard (1902). *Das Buch der Berufe: Ein Führer und Berater bei der Berufswahl; VIII. Der Journalist.* Hannover: Verlag der Gebrüder Jänecke.

Journalistischer Führer: Mitteilungen über die technische und redaktionelle Herstellung einer Zeitung (1999). Görlitz: Kommissionsverlag der Buchhandlug von R. Worbs.

Kaukiainen, Yrjö (2001). Shrinking the World: Improvements in the Speed of Information Transmission, c. 1820–1870. *European Review of Economic History*, 5, 1–29.

Keiter, Heinrich (1905). *Praktische Winke für Schriftsteller und Zeitungs-Korrespondenten: Mit den neuen Gesetzen über das Urheberrecht und das Verlagsrecht.* Essen: Verlag von Fredebeul und Koenen.

Kellogg, Ronald T. (1994). *The Psychology of Writing.* New York, Oxford: Oxford University Press.

Lieberwirth, Steffen (ed.) (2000). *Mitteldeutscher Rundfunk: Radio Geschichte(n).* Altenburg: Verlag Klaus-Jürgen Kamprad.

Lipartito, Kenneth (2003). Picturephone and the Information Age: The Social Meaning of a Failure. *Technology and Culture*, 44 (1), 50–81.

Livingston, Steven, and Douglas A. Van Belle (2005). The Effects of Satellite Technology on Newsgathering from Remote Locations. *Political Communication*, 22, 45–62.

Maibohm, Ludwig (1948). *Mit Schreibmaschine und Mikrofon … Aus dem Reisetagebuch eines Sportreporters.* Offenbach: Bollwerk-Verlag Karl Drott.

Mast, Claudia (1983). "Neue Technik"—neuer Journalismus? Zur Akzeptanz und Auswirkung des Einsatzes moderner Technologien. *Publizistik*, 28, 230–238.

Meier, Klaus (2002). *Internet-Journalismus.* Konstanz: UVK.

Müller, Friedrich (1900). *Schreibmaschinen und Schriften-Vervielfältigung.* Berlin: Verlag der Papierzeitung.

Nürnberger, Albrecht (1993). *Datenbanken und Recherche: Ein Handbuch für Journalisten und Dokumentaristen.* Remagen-Rolandseck: Rommerskirchen.

Peters, Dorothea (2007). Die Welt im Raster—Georg Meisenbach und der lange Weg zur gedruckten Photographie. In Alexander Gall (ed.). *Konstruieren, Kommunizieren, Präsentieren—Bilder von Wissenschaft und Technik,* 181–244. Göttingen: Wallstein.

Pöttker, Horst (2005). The News Pyramid and Its Origin from the American Journalism in the 19th Century: A Professional Approach and an Empirical Inquiry. In Svennik Høyer and Horst Pöttker (eds.). *Diffusion of the News Paradigm 1850–2000,* 51–64. Göteborg: Nordicom.

Prott, Jürgen (1983). *Berufsbild der Journalisten im Wandel? Zeitungsredakteure unter den Bedingungen der Bildschirmarbeit.* Frankfurt: R. G. Fischer.

Ratzke, Dietrich (1982). *Handbuch der Neuen Medien.* Stuttgart: Deutsche Verlags-Anstalt.

Rammert, Werner (1990). Telefon und Kommunikationskultur: Akzeptanz und Diffusion einer Technik im Vier-Länder-Vergleich. *Kölner Zeitschrift für Soziologie und Sozialpsychologie*, 42, 20–40.

Reinemann, Carsten (2003). *Medienmacher als Mediennutzer. Kommunikations- und Einflussstrukturen im politischen Journalismus der Gegenwart.* Köln: Böhlau.

Renner, Karl Nikolaus (2007). *Fernsehjournalismus. Entwurf einer Theorie des kommunikativen Handelns.* Konstanz: UVK.

Ropohl, Günter (2006). Der Begriff "Technik." In Gerhard Banse et al. (eds.). *Erkennen und Gestalten: Eine Theorie der Technikwissenschaften*, 44–48. Berlin: edition sigma.

Rühl, Manfred (2000). Technik und ihre publizistische Karriere. In: Otfried Jarren, Gerd G. Kooper and Gabriele Toepser-Ziegert (eds.). *Zeitung – Medium mit Vergangenheit und Zukunft. Eine Bestandsaufnahme*, 93–104. München: K. G. Saur.

Salcetti, Marianne (1995). The Emergence of the Reporter: Mechanization and the Devaluation of Editorial Workers. In Hanno Hardt and Bonnie Brennen (eds.). *Newsworkers: Towards a History of the Rank and File*, 48–74. Minneapolis, London: University of Minnesota Press.

Schiffhauer, Nils (2009). Das Schreibzeug arbeitet mit an den Gedanken. *Frankfurter Allgemeine Sonntagszeitung*, July 30, V8.

Schröder, Fritz (1941). *Ottmar Mergenthaler: Leben und Schaffen eines großen deutschen Erfinders im Ausland.* Berlin: Mergenthaler Setzmaschinen Fabrik GmbH.

Schütz, Walter J. (1963). Lochkartentechnik in der Pressedokumentation. *Der Archivar*, 16, 259–368.

Schulz, Klaus-Werner (1987). *Publizistik- und Kommunikationswissenschaft: Von der unsystematischen Handablage zur computergesteuerten Pressedatenbank: Bedingungen, Ziele, Chancen und Risiken des Strukturwandels von Dokumentationseinrichtungen im Medienbereich.* Baden-Baden: Nomos.

Te Heesen, Anke (ed.) (2002). *Cut and Paste um 1900: Der Zeitungsausschnitt in den Wissenschaften* (Kaleidoskopien 4). Berlin: Vice Versa.

Vaupel, Elisabeth (1987). Vom Gänsekiel zur Stahlfeder. *Historische Bürowelt*, 17, 5–9.

Wehle, J. H. (1883). *Die Zeitung: Ihre Gestalt und Technik.* Wien: H. Hartlebens Verlag.

Wilke, Jürgen (2004a). Die Telegraphischen Depeschen des Wolff'schen Telegraphischen Büros. *Publizistik*, 49, 125–151. Also in Wilke, Jürgen (2009). *Massenmedien und Journalismus in Geschichte und Gegenwart: Gesammelte Studien*, 232–260. Bremen: edition lumière.

Wilke, Jürgen (2004b). Vom stationären zum mobilen Rezipienten: Entfesselung der Kommunikation von Raum und Zeit—Symptom fortschreitender Medialisierung. *Jahrbuch für Kommunikationsgeschichte*, 6, 1–55. Also in Wilke, Jürgen (2011). *Von der frühen Zeitung zur Medialisierung: Gesammelte Studien II*, 315–376. Bremen: edition lumière.

Wilke, Jürgen (2010). Korrespondenten und geschriebene Zeitungen. In Johannes Arndt and Esther-Beate Körber (eds.). *Das Mediensystem im Alten Reich der Frühen Neuzeit (1600–1750)*, 59–72. Göttingen: Vandenhoeck & Ruprecht.

The Interpretive Turn in News

Kevin Barnhurst

On a Wednesday afternoon, May 6, 2009, a U.S. Senate subcommittee on communications convened on the Future of Journalism. Sen. John Kerry began with a welcome to the brave new world where "newspapers look like an endangered species." After comments by senators and testimony from press executives and others, online *Huffington Post* founding editor Arianna Huffington criticized the old system of journalism, saying it failed to explain "the run-up to the war in Iraq and the financial meltdown" but instead let "publishing pooh-bahs" dictate "what is important." Those days "are over," she said, "and thank goodness." The legislators asked whether online editions can produce reporting that explains events in depth, and newspaper publishers said no. When asked whether not-for-profits could fill the gap, a foundation witness said no. Sen. Kerry closed by noting that all sides agree on the need to preserve a kind of in-depth reporting that goes into *quality* news.

Scholars usually complain that U.S. news falls short when it comes to making sense of events, especially compared to the press in Europe. An article on "the collapse of journalism" in *The Nation* argues that other countries spend more on informing the public through media, an investment that yields "dramatically more detailed and incisive" news (April 6, 2009, 18). Another commentary points to the French press as an example of analysis and depth for American journalists to follow (*PressThink,* December 5, 2003), and a study of the Finnish press found an increase since the 1990s in journalists commenting on the front page about events (Väliverronen and Kunelius 2012). But sociologists report that U.S. local news fails "to produce reasonably unambiguous explanations"; and "answering the 'why' questions," they conclude, "might have attracted" a larger audience (Adkins-Covert et al. 2000, 241–242).

Critics have called regularly for more interpretive news. The left worries about commercial and public relations influences making news superficial, and the right about journalists' biases making news liberal, but both sides call for news that gives more context (Stauber and Rampton 2002; Weaver 1994). They say the press should do a better job of explaining where information comes from. Journalists also want to supply more and better interpretations. In recent decades they have decried a growing tendency of science news reports to manipulate facts (Crossen 1994), called for more context that "makes the complex coherent and meaningful," and warned against surrendering "their functions of analysis and explanation" (*Times-Picayune,* December 28, 1997, A10). Pulitzer Prize-winner Jack Fuller argued as publisher of the *Chicago Tribune* that audiences find "perfectly neutral accounts" uninteresting and expect journalists to do "much of the analytic work for them" (Fuller 1996, 30), but later wondered where the public would find adequate explanation (Fuller 2010).

The usual remedy scholars, critics, and journalists prescribe is for reporters to include more context and background in their stories, but not more judgment, which belongs on the editorial pages. The crisis of newspapers makes a remedy seem urgent. But was the news failing to supply citizens with the fifth W, the *why* interpretations they need to understand events? Did the content of news stories lack explanations of *how* events occurred? A closer look shows a broad interpretive turn in news, with explanations along with judgments and opinions increasing in daily newspapers, network television, public radio, and mainstream news websites.

The Trend in Newspapers

Since emerging as a distinct occupation, journalism has had an explanatory urge. The November 21, 1885, issue of *The Journalist* announced "Explanatory Journalism" on its cover page. In the twentieth century, newspapers expanded their emphasis on interpretations overall and their references to how and why news occurs (Barnhurst and Mutz 1997; Fig. 1). Compared to a one-paragraph story on a factory adding jobs in 1894, a 2005 story on layoffs at IBM runs twenty paragraphs (about average for *New York Times* staff pieces that year) analyzing global markets where jobs migrate to places like India and using IBM as "a corporate laboratory that highlights the trend." Reporting the event triggered the larger activity of interpretation.

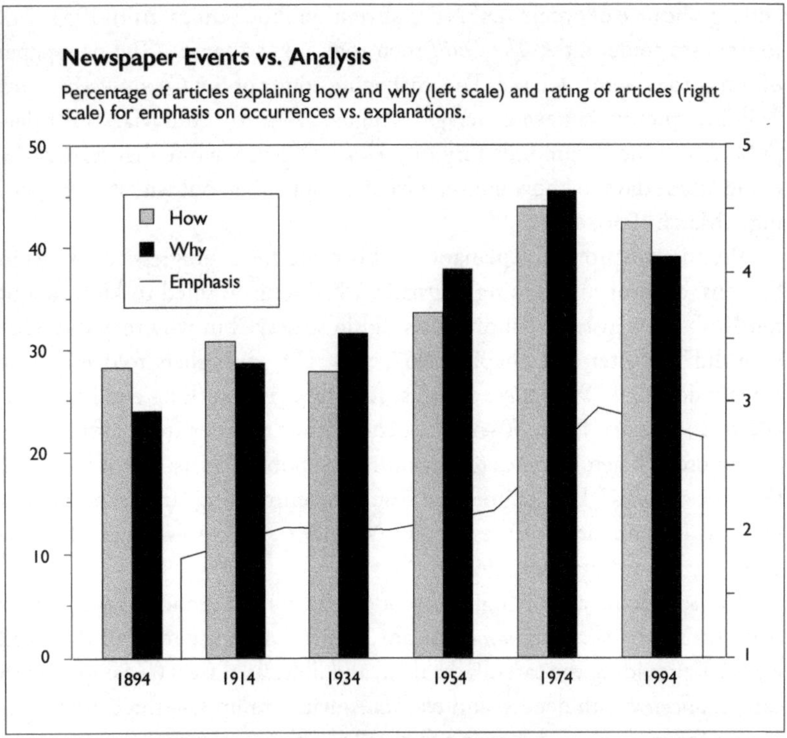

Fig. 1: Newspaper Events vs. Analysis
(Source: Barnhurst and Mutz 1997)

Stories without interpretations began to disappear as newspapers confronted competition from radio and the newly invented news magazine. Briton Hadden and Henry Luce founded *Time* in 1923, and the Mellon family and other investors in the 1930s backed a former *Time* foreign editor to found what became *Newsweek* (Wilner 2006; Brinkley 2010). A third competitor, *U.S. News & World Report,* started as a newspaper in the 1930s, and all three provided something newspapers lacked: summaries that made sense of the week's events. Radio news commentators had also begun drawing large audiences, creating pressure for the newspaper industry.

The American Society of Newspaper Editors in 1933 resolved, "Editors should devote a larger amount of attention and space to explanatory and interpretive news, and to presenting a background of information which will enable the average reader more adequately to understand the movement and significance of events" (ASNE Convention Proceedings, April 1933, 74). Journalism trade outlet *The Quill* then took up the cause. "The newspaper of tomorrow," wrote Charles Poe, managing editor of the *Chattanooga News,* "will interpret in its news coverage" (August 1934, 10). And Mark Ethridge, publisher of the Richmond, Virginia, *Times-Dispatch,* wrote that readers demand "these days to know not only what is happening, but why it is happening" (March 1936, 4).

Pressure to provide explanations also came from progressives. Lincoln Steffens, dean of muckraking journalists, had long "wanted to know about motives"—not just what individuals did in society "but why they did what they did" (Winter and Shapiro 1962, xiii). U.S. publishers told Raoul de Roussy de Sales: "We *believe* in facts, […] facts and nothing else but facts" (*Atlantic*, January 1937, 30–31), but the former reporter from leftist *Paris-Soir* wrote, "When it comes to informing the public," facts "are of less value than the analysis." U.S. editors had begun experimenting, and he pointed to the weekly summaries of the Sunday *New York Times* as evidence of a new tendency toward interpretation.

News presents an intelligible account by laying out and making events plain (*ex* means *out* and *planare* means level in Latin, the roots of the word *explain*), providing the careful, detailed, and illustrated facts (*expounded* with care, *explicated* with details, and *elucidated* with examples—three other synonyms for *explain*), and drawing them all together in the act of imagination and sympathy that goes below the surface to *interpret* the events. Although Webster's dictionary defines explanation as the central term, interpretation gathers together all related senses into an overarching term. The interpretive move in modern journalism tells the meaning of events, construed in light of general beliefs and judgments and particular interests. To interpret is to "represent by means of art," Webster says.

In the 1930s, "interpretative" news seemed admirable for attracting audiences, making sense of events, and doing public good. By 1939 a former Illinois editor noted, "The trend is unmistakably in the direction of combining the function of interpreter with that of reporter" and called "the ability to interpret […] a prerequisite" for "keeping the public informed" (MacDougall 1939, v). The Hutchins Commission report in 1947 reasoned that new technologies like radio and movies and more concentrated ownership had sparked a "communications revolution" (Commissions 1947, 4). To represent faithfully the disconnected groups of modern America to each other and to societies abroad, a responsible press must reject the "isolated fact" in favor of reporting "the truth about the fact" (ibid., 16, 22). Press responsibility means acting in the public interest by interpreting events and placing them in context, but the existing interpreters—columnists and commentators—focused on entertainment, deviance, and violence in an effort to reach larger audiences. The report says big business biased press content and advertisers dictated radio programming. Instead of representing society to itself, communications industries provided a "miscellaneous succession of stories and images, which have no relation to the typical lives of real people anywhere" (ibid., 58).

The ideal of social responsibility soon entered the college training of journalists. A 1949 textbook, *Exploring Journalism,* included a full chapter on Press Interpretation, where the two authors, a journalist and an academic, argue, "It is not enough for the public to have the facts, for it also must know the meaning of the facts," and a majority of journalists act responsibly when they "stress interpretative reporting" (Wolseley and Campbell 1949, 359, 360). The ideal became entrenched in journalism education once the Cold-War-era *Four Theories of the Press* placed social responsibility opposite totalitarian journalism. But the authors said journalists were "deficient in performing" their task of "enlightening the public" and should do more so that the people can self-govern in democracy (Siebert et al. 1956, 74).

Working reporters and news executives had qualms. In a 1954 collection, *Newsmen Speak,* journalists are unanimous about accuracy and about separating opinion from facts, but not about interpretations. Some say coverage "should go a step farther by explaining the news" (Coblentz 1954, 47), but a New York AP bureau chief calls for a facts-only news story "so written that the reader will understand it easily and completely" (ibid., 139). William Randolph Hearst, Jr., is lukewarm, calling for news to interpret "if necessary" (ibid., 74). But another publisher says reporters had moved "from scoop to scope" and were writing "with deepened perspective and understanding" (ibid., 191). A news agency president says his services provide "the best possible coverage" by reporting "news events themselves" along with "what the facts of a news story mean" to make them understandable "for the average citizen" (ibid., 148).

Despite objections in the journalism trade press, "interpretative" news did grow from the 1930s to the 1950s in our data. The magazine *American Journalism* voiced concern in 1941 that interpreting events slides quickly into expressing opinion; an early issue of *Nieman Reports* criticized editors in 1949 for letting too much opinion enter into interpretive stories; and the monthly *Bulletin* of the American Society of Newspaper Editors continued the controversy later (Landers 2005). But the turn toward interpretation continued amid concerns that new communication technologies of the period threatened a press deficient in explanations.

New & Other Journalisms

By the 1960s mainstream newspapers began shifting toward "explanatory" news. The owners of the *Wall Street Journal* in 1962 founded *The National Observer,* a weekly newspaper of interpretation that pioneered articles on trends and current event analysis, genres the daily press imitated. The *New York Herald Tribune,* despite its demise in 1966, provided another model of explanatory journalism under former *Newsweek* editor John Denson (Landers 2005).

Movements from beyond American journalism and from its fringes pressed for more explanatory news. In the heady days of the 1960s, journalists and academics shared a faith in the potential for poor countries to advance through the miracles of agricultural innovation. Journalism seemed useful to help transform economic conditions, leading to individual freedom and democracy (Ogan 1982). In the Philippines, agriculture extension communicators ran a successful media campaign to inform small farmers about new rice varieties (Shafer 1991). A press institute and later an academic program in the Philippines founded *development journalism,* meant to boost national advancement through explaining the Green Revolution and economic change. Similar efforts emerged in India, and seminars of the Press Foundation of Asia trained journalists to interpret development projects, rather than merely passing on government claims. U.S. journalists saw no ready application at home and raised objections when governments abroad co-opted development journalism to restrict press freedom, which the annual Associated Press survey found "seemed to lose more ground than it gained" around the globe in 1966 (*New York Times,* January 2, 17).

Another movement to explain and interpret news in the 1960s, the alternative press, drew inspiration from an older model, the *Village Voice* (McAuliffe 1978). Its journalism combined reviews of art, movies, and theatre with rambling personal stories about local events. The weekly did no editing and paid writers little or nothing at all. Alternative newspapers used methods from mainstream journalism to attack it and foster activist writing about the counterculture. But within a decade the mainstream press had co-opted radical weeklies by expanding coverage and explaining youth culture. The surviving alternative papers in major cities followed standard business models, like the *Voice,* but continued to challenge objective writing and push for explanatory articles.

The *Voice* was a model for New Journalism, which burst on the scene by the mid 1960s. Reportage by Norman Mailer and others for *Esquire* magazine inspired writers to experiment, and Truman Capote's *In Cold Blood* became a classic of the genre because its interpretation followed facts. Reporters in the daily press began doing interpretive writing: Tom Wolfe at the New York *Herald Tribune,* Gay Talese for the *New York Times,* and others. But the journalism profession had doubts. A former *New York Times* editor issued a screed (his word) saying New Journalism was not new, not journalism, and "cannot and should not take the place of the misnamed 'Old Journalism'" that critics scorned as the "who-what-when-where-how-and-why school" (*ASNE Bulletin,* January 1972, 6). Even if 'new journalism' strained credibility, old journalism was "increasing [...] interpretation—the effort to make the news clear and relevant for the reader" (ibid., 8–9).

As interpretive news advanced, the American right wing took up press bashing. During the 1964 National Convention, when Dwight D. Eisenhower urged Republicans not to let outsiders from the press divide them, the delegates directed "a deafening roar of boos at the press stands flanking the speakers' platform and many on the convention floor jumped up and shook their fists" (Wicker 1978, 1–4). Tom Wicker, a *New York Times* staffer, remembers "those shouting, livid delegates, rising almost as one man, pointing, cursing, in some cases shaking their fists, not just at the men in the glass [television] booths but at *me.*" He sat with other "reporters gazing in astonishment at this sudden surge of hatred." His paper could not report what happened, he says, because the account would have introduced subjective explanation. But the era had begun when conservatives saw the press as a powerful agent of "menacing, hateful, corrupt" interpretation (Wicker 1978, 16).

Mainstream reporting on youth culture and protests made major daily newspapers an easy target as the radical press went into decline. The advocacy *I. F. Stone's weekly,* which predicted and then protested the Vietnam conflict, had 70,000 subscribers when it closed in 1971, but analysis suggested all journalism would move toward interpretation (Murphy 1974). Chicago *Sun-Times* and *Associated Press* reporting illustrates the spreading influence of immersion experiences and literary writing to interpret them. In the sixth edition of his textbook, MacDougall lists "activist, advocacy, participatory, [...] investigative," and "reformist" among the new kinds of journalism expanding interpretation in the press (MacDougall 1972, v). New Journalism continued the long connection between news and literature (Fishkin 1985), extending how journalism "conveys moments in time" and "explores how and why" (Connery 1992, 5). By the mid 1970s, when interpretation had become widespread in American news (Johnstone et al. 1976), mentions of how and why events occurred and emphasis on explanations reached a zenith in newspapers, just as movements to make news more explanatory pressed on mainstream journalism from all sides.

Interpretive Movements after Midcentury

Practitioners and academics pushed other interpretive reforms for journalism. Peace journalism had origins in the pacifist societies of the early nineteenth century, but formed part of peace initiatives in the 1970s (Lee and Maslog 2005). Peace instruction entered U.S. colleges of the Vietnam War era, sometimes at the urging of students who had studied journalism. In the build-up to subsequent wars, news work routines failed to provide the context for the public to judge events (Machin and Niblock 2006). Journalists in surveys agreed that explaining was their greatest challenge, and some reforms grew from practices such as *precision journalism,* the application of data analysis to news (Meyer 1973). The techniques soon grew common for criminal cases, election claims, drug trafficking, and civic engagement, especially among journalists dissatisfied with traditional tools of interpretation (Mosier and Ahlgren 1981).

Investigative journalism fired the imaginations of journalists and the public with how news could change public affairs. Its early roots were in muckraking and crusading news, but the 1970s movement was mainstream (Armao 2000). The *Plain Dealer, Miami Herald,* and *Newsday* set up investigative beats, and the *Boston Globe* and other papers followed suit. Investigative journalism gained national attention in 1969, when atrocities in My Lai, Vietnam, came to light, and after investigations exposed crooked fundraising at the Nebraska Boys Town orphanage and documented the harsher sentences that black and poor defendants faced in Philadelphia. Richard Nixon resigned the presidency in 1974 once other institutions intervened, but the

Washington Post Watergate investigation and media coverage kept the scandal alive (Lang and Lang 1983; Schudson 1992). Skepticism grew toward government, and a romantic notion of journalism arose. Schools soon produced a glut of journalism graduates, and a mainline church then helped organize *Investigative Reporters and Editors*, its acronym IRE expressing a moral stance (Houston 2010). Critics objected that investigative reporting tended not to focus on corporations, advertisers, or marginal groups and made journalism political, aiming to expand beyond social into moral responsibility that usurped "the functions of institutions" (de Burgh 2000, 84). The new interpretive movement was open to abuse, but in its heyday even government officials criticized the press for not doing enough investigating.

Investigative, precision, and peace journalism were exceptional, a few reporters usually working on their own time and without much support from skeptical management. Even in boom years, conditions always worked against them: corporate ownership, market competition, government deregulation, information-as-commodity, and underfunded original reporting (de Burgh 2000). But interpretation had become basic to news writing by the 1980s (Weaver and Wilhoit 1991), and in 1985 the Pulitzer board began awarding a prize for explanatory work. In its first quarter century, the winners were primarily complex stories that explained the human consequences of medicine and science, economics, and government policies.

But the top-down interpretive pattern permeating American news from beginner training to Pulitzers Prizes helped give rise to civic or public journalism. After journalists interpreted the 1986 and 1988 elections as a game of insider jockeying and failed to counter manipulative political ads, public journalism began to emerge mostly at small heartland newspapers (Rosen 1999). To reengage citizens facing too much information, journalists again felt restless with just-the-facts news. Within a decade more than two hundred projects sprang up, got attention, and began influencing mainstream practices. But heavy hitters from the elite press, especially the *New York Times,* soon condemned the idea (Corrigan 1999). Critics said public journalism was doing only what journalists should be doing anyway,—describing and explaining events—but in ways that distorted newsroom budgets and compromised press independence.

American news became more interpretive as waves of reform demanded: "interpretative" news of the 1930s; New Journalism of the 1960s; the surge of peace, precision, and investigative work in the 1970s; and public journalism arising in the late 1980s. The prominent rejection of each movement denied the changes underway, and interpretations declined somewhat in news in the late 1980s and early 1990s. But the interpretive turn rebounded over the next two decades. Despite persistent complaints that the news explained too little, journalists on the ground were doing the opposite. They became interpreters of public life.

Television in Context

Newspaper journalists said they had to make their stories more explanatory after "the nightly 'front page' on television" (Wicker 1978, 16) began leaving them to write follow-up coverage. TV images can show events without the delays of writing and publishing, and critics soon lamented that broadcasting lacked interpretation. In 1980 *New Yorker* staff writer George W. S. Trow complained about no-context television events (November 17), and again reproached journalists twenty years later for failing to connect the dots: "television news people," he writes, "value and love the episodic" events left to stand without interpretation (Trow 1999, 44). Critics since then agree that television fails to supply "the context and analysis necessary to turn facts into real understanding" (*Boston Globe,* April 7, 2003, A19). In 2010 another *New Yorker* critic saw further declines in reporters "seeking context" (*New Yorker,* January 25, 2010, 42).

The American public has tended to view problems as "concrete rather than abstract" and rely increasingly on television for news (Iyengar 1991, 136). Episodic newscasts may lead audiences to ignore the social conditions and public policy behind problems. For a century the public has scored poorly on political knowledge, making it unclear how democracy is possible, especially when television adds little to what citizens know (Barnhurst 2012). An uninformed audience needs more explanations, but did the interpretive turn fail to spread to television as critics suggest? No, journalists speaking on the national U.S. evening news have expanded their interpretations since the 1960s (Steele and Barnhurst 1996). After scooping newspapers on breaking stories, network newscasts themselves began shifting into interpretive styles instead of sticking with episodic coverage (Fig. 2).

In 1968 political reports on national television gave a direct description of events, some with film segments of candidates talking. With each passing election, candidates responded to the visual rhythms of television news, and

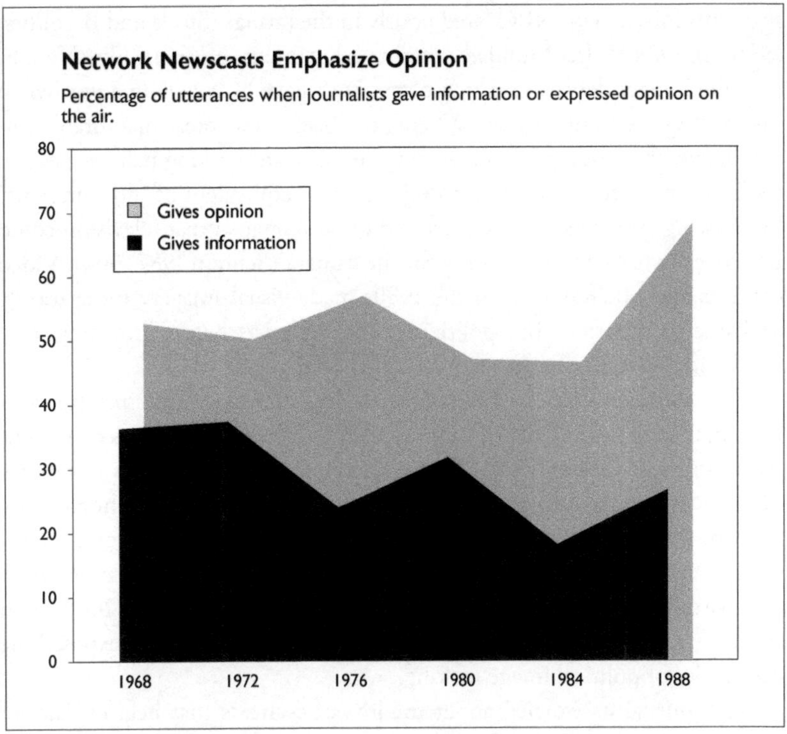

Fig. 2: Network Newscasts Emphasize Opinion (Source: Steele and Barnhurst 1996)

journalists delivered more opinion (Steele and Barnhurst 1996). In a 1972 news package, the NBC anchor's opening is neutral and informational, as is the correspondent's first voice-over, and after a sound bite of George McGovern's stump speech, the correspondent adds some context, explaining that McGovern may give a national speech on television because "his image is still blurred among many regular Democrats" (Steele and Barnhurst 1996, 199). Television news gained dominance in the 1970s by limiting politicians' sound bites, deferring less to campaigns on what to show, and interpreting candidates' appearances (Henry 1985). In 1980 Ronald Reagan countered by issuing pithy made-for-television phrases on eye-catching backdrops that his opponents tried to emulate.

The evening news responded by shortening sound bites further, being more selective about images, and trying to explain "the issues," although the segments initiated on NBC fared poorly in the ratings (Steele and Barnhurst 1996). In 1984 Walter Mondale presented documents on an un-filmable subject, the budget deficit, to which newscasters gave little play in a year when his most quoted phrase was "Where's the beef?" and most memorable image showed him riding a tank wearing an awkward-looking helmet. Reagan took the opposite approach, refining the visual equivalent of "no comment" by showing reporters how he preferred to be shown so that television could not cover "what he wasn't" doing for the camera (Schram 1987, 304). Video cameras and satellite feeds in the 1980s made visual imagery more readily available to reporters, but politicians and lobbyists began providing local news with video press releases or so-called "wallpaper."

Campaign coverage by 1988 shows a clear increase in interpretive journalism, mostly in the form of opinion. A report from ABC News condemns the motives of George H. W. Bush, who appears in shirtsleeves to portray himself "as a man in tune with rural America" (Steele and Barnhurst 1996, 189), but rides a luxury tour bus and limits press access. Hampered by declining audiences, staff, and budget, the networks cut back on making their own visuals, but image-consciousness held sway (Adatto 1993/2008), and so they treated campaign generated images with skepticism and exposed the mechanics of political image making.

But journalists worried about the loss of contexts that help explain an event's meaning and called the new visual style of network news "MTV journalism, images with little context" (*New York Times,* December 28, 1989, C18). The networks did show more images at a much faster pace, backdrops became more varied, and for the first time journalists appeared most often on location in the 1990s (Barnhurst and Steele 1997). In 1992 the candidates tried to circumvent the national networks (Goldman et al. 1994). Bush worked through local news and satellite feeds to small groups and individual stations. Independent Ross Perot used "infomercials" and went on *Larry King Live.* Bill Clinton found alternatives like *Saturday Night Live* and set a daily agenda to dominate the "media spin." Through imagery, the networks found "context" could grab viewers and keep them watching, but instead of documenting events as they unfolded, the images were a backdrop for journalists, who appeared more frequently and closer up, as the center of coverage.

The new visual style of news reports treated the public as passive observers, supplying images that tended to work against learning. Although rich in information, the routine scenes could show little "about complex social and political problems" and instead reinforced stereotypes (Graber 1990, 139). As context, visuals added variety to the public airwaves, but may have impeded careful thought or reflection among viewers. By the 2000s, newscasts had abandoned their early descriptive style, but interpretive reporting "did not translate into higher-quality analysis" (Patterson 2003, 7). After facing widespread criticism for calling the 2000 election winner too early, newscasters promised changes, but in 2004 they continued their "journalism of assertion" (Marriott 2007, 699), despite some changes in handling exit poll data.

As television entered American politics, candidates and newscasters competed to capture voters' imaginations through interpretive visual styles. Candidates influenced journalists, and through them the public, by staying on message and managing imagery. But journalists influenced the public and policymakers by analyzing events, judging politicians and their proposals, and supplying more interpretive news (Barnhurst 2012), activities that have grown in TV news coverage. The interpretations put journalists front and center, showing disdain for the managed images and words coming from the campaigns.

The most watched but lower status segment of local news, the weather shows how deeply the interpretive turn affected newscasts (Doherty and Barnhurst 2009). TV forecasters have long been the butt of jokes when not joking themselves, but when Boston hosted a political convention, the morning weather forecasts were serious about the national political ritual. The chitchat among announcers and other on-air personalities and the visuals during report segments on three network affiliates illustrate how weathercasters acted as explainers for the public, combining complex visual with narrative elements to heighten drama. They featured themselves at times—and surprisingly—as controllers of the capricious weather that might threaten the political convention. The public rarely figured in weather segments and only in passing during news team banter. The interpretive turn had reached the margins of journalism.

Television journalists expressed opinions more than print journalists did, but both media added opinions as critics feared. Journalists said they were responding to pressures from markets and new technologies, a pattern deserving a closer look. But the interpretive news of assertion shifted attention to journalism itself, especially on television. Was the stronger move toward opinion particular to television, with its reliance on imagery, or general to broadcasting?

The Case of National Public Radio

Radio seemed likely to give the answer because it uses only mental imagery. It also might have bucked the trends in newspapers and television. Radio had moved into commentary early in the twentieth century and had an honored tradition of mainstream explanatory news going back to Edward R. Murrow's CBS radio reports from London during the German bombing blitz of World War II (Seib 2007). By the 1980s, U.S. commercial radio had moved to narrower music formats with less news, but National Public Radio became a home for pieces interpreting events. Did *Morning Edition* and *All Things Considered,* founded as outlets for extended audio reporting, follow the interpretive turn? And did they rely on opinions as television did?

Given the history of radio, it came as no surprise that reports on NPR in 1980—the first election year both programs were on the air—began with more interpretation than newspaper articles had (Barnhurst 2003; Barnhurst and Mutz 1997). But NPR reports became even more interpretive (Fig. 3). That emphasis peaked in 1996 and by 2000 was a third higher in than where it began, changing over a relatively short period.

In 1980 a report on some Jimmy Carter hecklers opened with facts, and any terms expressing an opinion, such as *mincing* and *defect,* came from the candidate. In 1996, a year when news interpretation reached a peak, a report on the presidential debate and health care opened with the big-picture interjection, "What a difference four years makes!" (Barnhurst 2003, 10). Then after an archival sound bite of Clinton came the reporter's judgment of that his health care proposals were "vaguer, less-ambitious" than in the past.

As the interpretive turn advanced on NPR, reporters began interviewing each other (Barnhurst 1991). In a 1996 segment, Bob Edwards asks, "So, do you think—was [candidate Bob] Dole's move into the ethics area too tepid?" And Cokie Roberts responds, "I don't know how he could have done more," both journalists speaking from their own expertise (Barnhurst 2003, 17). Similar examples became more common, with NPR journalists in the role of sense-makers. Asking questions, an archetypal task for journalists, declined sharply as journalists paraphrased their sources instead of airing questions and answers. The mainstay of giving information also seesawed down, so that by 2000 it no longer represented the largest share of speech, and journalists expressed opinions more than twice as often. In the hotly contested election that year, NPR journalists focused less on action. More than half the times they spoke, they were interpreting. As in network TV newscasts, how

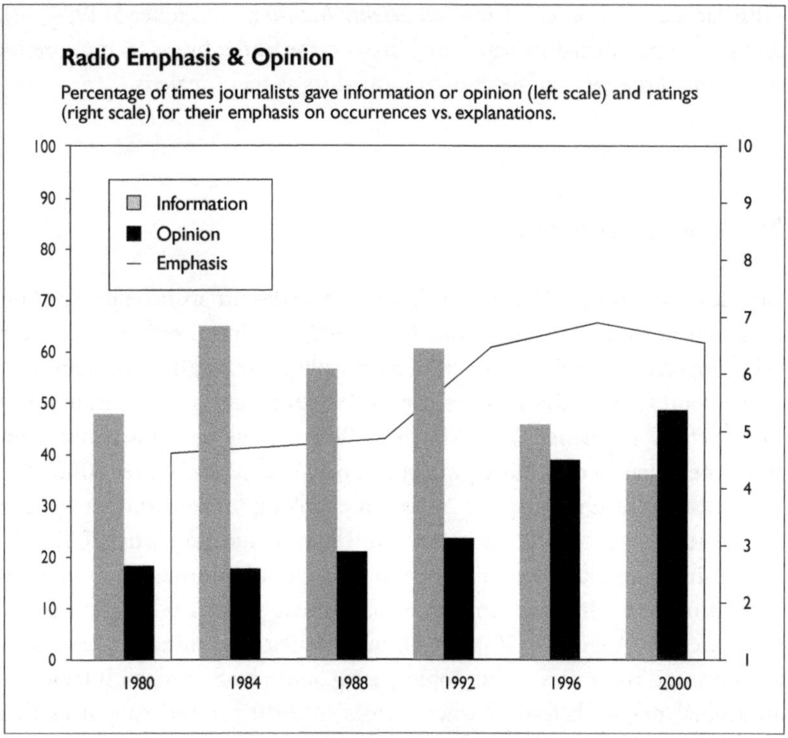

Fig. 3: Radio Emphasis and Opinion
(Source: Barnhurst 2003)

NPR journalists explained the meaning of events shifted, but not to providing more of the context; the share of times they gave background declined by more than half.

NPR moved into the mainstream (McCourt 1999), becoming more like the daily press and network television. As it grew prominent and attracted larger audiences, its programs expanded their political reporting, a premier beat for reporters and marker of authority. Reports became less neutral in tone and more polarized, making the news more dramatic for listeners but also a target for critics and for attacks on its funding. Like television news, NPR relied on more opinions to interpret events, a tendency common to the interpretive turn but more pronounced in broadcasting. And NPR also faced charges that its reports failed to give *enough* of the context for events, especially in the 1990s after Congress cut the network budget amid charges that

NPR lacked balance (e.g. *Christian Science Monitor,* December 5, 1994, 18). As earlier elite journalists had feared, news came under fire as its interpretive style shifted the focus of news onto journalists as sense-makers.

News on the Internet

Broadcast and other industry insiders say that, despite its benefits, the "Internet Hurts Journalism" (*Atlantic,* April 2009). Their expressions of concern echo responses to earlier shifts in news distribution. Awash in data without much contextual analysis, voters cannot "separate reported fact from fabricated fiction" (*Television Week,* March 1, 2004, 18), say journalists, and so internet news has "a credibility problem" (*Nieman Reports,* Winter 2006, 37). Journalists in turn must spend "more time looking for something to add to the existing news, usually interpretation" (Kovach and Rosenstiel 2001, 77).

Critics agree that the public encounters a glut of information online and finds it hard to "sift fact from rumor or evaluate sources or gauge context" (*Boston Globe,* August 18, 2008, A15). Interpretive news shifts emphasis onto the surrounding context, a hot topic during South by Southwest Interactive, an annual new-tech festival where critics in 2010 reached consensus that journalism online should use "*topics* instead of stories as the atom of news" (SXSW, Austin, TX, March 2010). Besides linking related content, topics are interpretive, panning away from events to show the context—how and why they happen. Researchers noted the rise of interpretive news by the late 1980s (Barnhurst 1991) and agreed two decades later that it is "one of the key concepts" in "the style and character" of journalism (Salgado and Strömbäck 2012, 144). As mainstream news became more interpretive, journalists became direct participants "no longer constrained by a need to place newsmakers' words and actions at the center of the story" (Patterson 2000, 14). Context had become king.

Did news grow more interpretive as it moved online? In websites of the same newspapers analyzed from 1894 to 1994 (Barnhurst and Mutz 1997) the tendency toward interpretation had increased by 2001 (Barnhurst 2002), and then after dropping off through 2005 (Barnhurst 2010), it reached a new peak in 2010 (Barnhurst 2012), running in line with the historical trend (Fig. 4). The reporting on the websites—still in the lower half of the scale—was

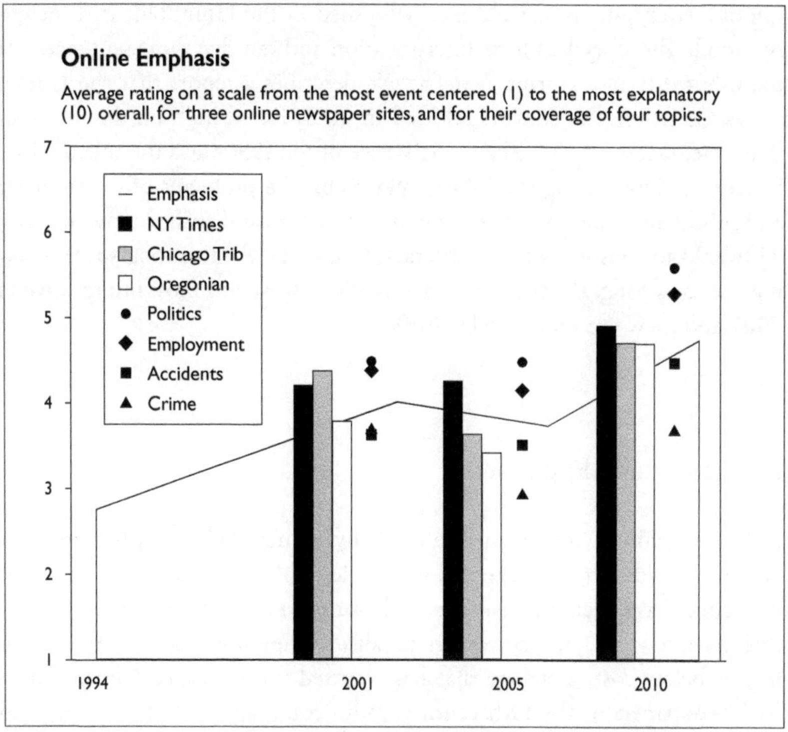

Fig. 4: Online Emphasis
(Sources: Barnhurst and Mutz 1997; Barnhurst 2009, 2010 and 2011)

approaching the tendencies in media known for news analysis, such as NPR (Barnhurst 2003).

Explanatory tendencies grew consistently at the *Times,* even while dipping temporarily at the other two news sites, especially at the *Oregonian.* The articles came from news columns, not the opinion or editorial pages, and rarely carried the label "News Analysis." The four topics—politics, employment, crime, and accidents—aligned predictably, with stories about politics and employment including much more interpretation than those about crime and accidents. Topics also spread out, so that accident stories ended somewhat lower by 2010. As news stories became longer, they tended to be more interpretive.

Two examples can illustrate the range from short and factual to long and interpretive. A *Tribune* story says police are investigating the drive-by shooting of a twenty-four-year-old man who died in the Humboldt Park neighborhood. The story had little interpretation and ran five short sentences. At the other extreme, a print *Times* article describes a report that the United States Conference of Mayors released in 2001. Rated high on the interpretive scale, it reviews the urban renaissance of the 1980s and the role of cities in creating jobs during the 1990s, talks about the problems of urban decay and suburban sprawl, and cites economic forecasts and authoritative sources, with links to census maps and interactive features. All this took space, making the story long. Political reports were the longest and most interpretive in 2005 and grew even more so by 2010.

Explaining the Explainers

How to explain what happened to the "why" of news? One way is to look for patterns. Insiders had a patterned way of describing changes in the industry. Audiences have been increasing recently for news content, but executives say they have fewer paying customers because competitors distribute news for free or below cost, a change that has recurred for a century. Cheap industrial newspapers of the 19th century, radio commentators of the early and television newscasts of the mid-20th century, and online news aggregators and blogs of the 21st century undercut older news outlets by using newer technologies. The alternatives might seem less expensive or more convenient for the public, but journalists argued that the shifts debased news content. Newer outlets tended to reuse facts from existing media, add opinions, and pursue sensationalism. Editors in each period described the change as a crisis for public life. The cheap outlets using new techniques but leaching from or undermining the mainstream media, say industry insiders in times of change, threaten access to information citizens need in democracy.

As each new mode for transmitting news emerged, the content followed a surprising pattern as journalists responded to different pressures by doing more of the same. When news magazines and radio commentaries began supplying more interpretations, newspaper journalists said they had to *do the same* to compete. After mid-century, television news supplied more breaking news of events, and print journalists saw themselves pushed to do the opposite of news flashes, that is, *more of the same* interpretive news they had already been adding, again to compete. In the new century, the internet began generating more interpretive opinions, and journalists at legacy media saw the need to *do more of the same* by sorting out and making even better sense of events, expanding interpretation. Across all periods, a perceived threat to mainstream news met with the same general response, a demand for more interpretive journalism and an expanded role for journalists as explainers.

In another pattern, prominent leaders in news organizations protested in each period, disparaging interpretative news of the 1930s, New Journalism of the 1960s, and public journalism of the 1990s. Whether or how much the boundary shifted between an ideal of factual news and the slippery slope of comment and opinion diverts attention from the interpretive trends. The leaders who objected were themselves proposing forms of interpretation. Each new journalism movement since the Pulitzer era aimed to tell the truth in ways that would protect the public and make for a better world. The journalists raising objections made fine distinctions without challenging the interpretive urge basic to U.S. journalism. In the face of critics who complained of a deficient press, mainstream editors, publishers, and managers called for more context, more background, more depth, and better analysis while cautioning that judgments and opinions can erode news credibility. But interpretation is inherent in the finding, selecting, and laying out of facts—in careful, detailed, and lucid description. Going below the surface and presenting an insider account are acts of imaginative interpretation. The project to make the world intelligible to the public is interpretive to the core.

The debates over interpretive news spread beyond practitioners, but the U.S. left and right swapped positions in the course of the past century. Reformers of the early 20th century saw fact-centered news as an antidote to the excesses of partisan journalism, but the ground had shifted after mid century, when convention delegates jeered at press boxes. The political right since then has called for what progressives on the left had once proposed. The former editor of *The American Scholar,* a neoconservative, charges the *New York Times* with a "yawningly predictable" liberal point of view that treats "the facts […] as so much tedious filler" in "so-called hard news" (*Commentary,* January 2006, 47–48). After reciting the claim of a liberal bias among journalists, he calls for a return to "integrity and impartiality" (ibid., 51), for facts based in reasoned argument about concrete reality.

Equally serious thought from the left about journalism tends to worry less about bias than superficiality. A *New Yorker* critic writes that the loss of "professional expertise" as newspapers cut reporting jobs leaves the public without what it needs to know, a gap that cheap "aggregation and opinion" cannot fill (*New Yorker,* March 31, 2008). Online options lack the staff and resources to replace mainstream newspapers, and peddling the party line is shallow whether at the *Huffington Post* or Fox News. Liberals in *The Nation* magazine argue for government to shore up credible news analysis and insight by providing subsidies that would reverse the "lowering quality and [...] trivializing" of journalism (January 7, 2010, 4, 6). Corporate ownership has either watered down of news content or replaced it with fluff (Bagdikian 2004), and a reliance on imagery makes television news especially vulnerable (Adatto 2008). The solution to the problem of superficiality, liberal critics say, is better explanations.

Reversing positions among left and right observers turned on questions of power. A bright line between news and opinion is difficult to draw at the fuzzy boundary where journalism moves beyond "superficial" facts into interpretive "bias." Over the past century, each new technology produced a crisis by threatening the authority of mainstream journalism. The response from the press consistently pushed toward greater interpretation, an activity of the powerful. The journalism leaders who protested recognized the political risks of looking like interpreters instead of reporters, but their protestations also attempted to redraw the line, a moving border that shifted polarities along with political positions on interpretive news.

Realism Found & Lost

Interpretive news grew from roots in twentieth-century realism, following movements in Europe centered on the reality of objects, words, and images. In the sciences realism assumes a world exists independent of any observer, and humans can either take the measure of objects or come up with theories that describe its unobservable but solid dimensions. Early in the twentieth century, when Albert Einstein won the Nobel Prize, his theory of relativity produced "a profound sensation" (*Washington Post,* January 18, 1920, 43; *New York Times,* December 3, 1919, 19, and *Washington Post,* December 14, 1919, E5), and a few years later quantum theory reached wide public attention through physicist Nils Bohr's series of lectures at Yale (*New York Times,* November 7, 1923, 9). The new science called into question commonplace ideas about a fixed reality at the macro, interplanetary scale, and also at the micro scale where atomic objects can at once behave as particles *and* as waves.

Fundamentalist religious leaders of the era responded to emerging science by reasserting direct or naïve realism (Marsden 2006), one in a long tradition of common sense ideas in American history (Rosenfeld 2011). In the sphere of objects, realist or objective views continued to pervade American culture, journalism, and even some fields of science for most of the twentieth century (Rudner 1953).

In language, American literary realism expressed the progressive reform urges of a century ago. Growing industrial production, migration into urban areas, and distribution of goods for mass consumption produced a chaotic new world, and so writers tried to help manage the complexity and control the confusion of rapid social change (Kaplan 1988). The new journalists of the time explained and interpreted conditions by presenting facts and narrating a closely observed reality, and novelists did something similar, some using a common trope of claiming their accounts were factual (Davis 1983). Realist writers hoped for progress against social injustice and pursued verifiable, objective truths using methods akin to science (Pizer 1998). In the realism of the period, journalists and novelists held out hope that the new world could be coherent, open, and democratic.

The visual sphere turned to realism with the rise of photography and then moving pictures. Photographers of the Progressive Era aimed to document the emerging reality of modern mass-scale industrial life and, by making the downtrodden visible, spur reforms to improve the lives of immigrants, workers, and ghetto dwellers. Photographic images seemed to interpret everyday life (Barnhurst 1996). Lewis Hine, who documented three decades of immigration, exposed child labor, and recorded the building of the Empire State Building, said that the modern visual imagination could bring life "to facts, to science, and to common lives" (Trachtenberg 1997, 132). The rise of the picture press and entry of halftone photographs into daily newspapers seemed to make readers witness to current events (Barnhurst and Nerone 2001). Philosopher John Dewey said the art of "presentation is fundamentally important" because "artists have always been the real purveyors of news," and democracy would flourish through "the art of full and moving communication" (Dewey 1927, 183–184). Only art was powerful enough to match the world emerging in the twentieth century.

Visual realism, the idea that optical lenses and camera technology can take a true measure of the world, persisted through most of the twentieth century. In the 1960s newspapers were already losing influence and closing down editions—consequences of their focus on entertainment instead of "real news" based on facts. Daniel Boorstin exposed the pseudo image, "the making of our illusions—'the news behind the news,'" aimed to "clear away the fog so we can face the world" (Boorstin 1961, 33, 5–6). But his critique reasserted realism. An art historian and visual theorist later called for an alternative that would accept the truth of the camera image as "an objective record" that "has to be interpreted" (Gombrich 1980, 238). His attempt to overturn commonsense realism—"copy theories" of representation—still proposed "a universal science" that excluded values, becoming again "just another realism" (Mitchell 1991, 25, 33).

Only later in the century did the solidity of realism show signs of cracking. For objects, the theories of relativity and quantum mechanics became commonplace. Newtonian science seemed less accurate when new ways of living alongside electronic devices defied time and place. For language, a less-sanguine view of realism emerged. One literary editor objected that the successors to realism simplified complex problems and reduced the social to the individual (Cowley 1947). In lectures during the early 1960s, essayist Mary McCarthy worried that a focus on facts, fairness, and fidelity—the "journalistic frame"—would ultimately condemn the modern novel, which "with its common sense, is of all forms the least adapted to encompass [a] world, whose leading characteristic is irreality" (McCarthy 1960, 433, 455–456), the same term Gombrich employed. Critics implicated American realism in the rise of consumer culture. Reporting on everyday life displayed events for consumption and brought consumers under surveillance, but also alleviated their fears of powerlessness under the new industrial order. Not "a progressive force exposing social conditions" (Kaplan 1988, 7), realism made consumer relations the norm. Mass circulation journalism—the new medium of the era—competed to do the same, promising "a coherent and a cohesive world" but delivering only insoluble problems (ibid., 13). Little wonder political positions changed sides! The language of modern realism lingered in journalism and other literary genres, but could not cope with the growing irreality, in McCarthy's term, that came to dominate U.S. American culture by the century's end.

For visuals, essayist Susan Sontag proposed in the 1970s that the camera and expert photography made the public into passive onlookers not activists (Sontag 1977). Science and government used photographs to monitor and control the populace, and politicians, celebrities, and pornographers used them to make mythic icons of themselves. Media used them as evidence to support dominant ideas about nature and society, and advertisers and marketers used them to immerse the public in a "commodity dream" about products and tourism that promised an escape from the mundane, industrial world (Berger 1980, 58). The ensuing critique of positivism—a philosophy of that makes sensory experience the source of authoritative knowledge (Orvell 1989)—accompanied the rise of digital picture manipulation (Perlmutter 2005). By century's end, objectivity seemed less neutral and solid (Haskell 2000), as the public became aware of photo manipulation and grew more skeptical or even cynical about visual images.

The Persistence of Realism

By the twenty-first century, public awareness had encompassed not just the solidity of things but the quarks and chaos of "postmodern science," not just the realism of language but post-realist literature, and not only the specificity of images but new forms of manipulated irreality in the camera arts, a terrain of contradictions (Merchant 2004, 212, 217; Swinford, 2001). But philosophers and thinkers in journalism had already planted the seeds of irrealism at the beginning of the twentieth century, just when realism established itself. Italian writer and political theorist Antonio Gramsci attacked the press for injecting the public with "ways of feeling and judging the facts" that would benefit the press itself and serve the dominant classes (*Avanti!*, December 22, 1916). Walter Lippmann, after echoing Gramsci on coverage of the working classes, wrote that "immediate realities lie outside the direct experience both of the reporter, and of the [...] public," making news is an "abstraction" of reality based on catchwords invested with feelings," which for them "are realities" (Lippmann 1922, 348–349). Argentine writer Jorge Luis Borges imagined interpretation as "labyrinth of symbols" where humankind would become lost (*Ellery Queen's Mystery Magazine,* August 1948, 107)—in today's terms a digital news site, massively multiplayer online game, or hypertext novel hosts a universe in utter contrast to the solidity of Newtonian physics.

U.S. journalism remained nominally realist. At mid-century neither news workers nor readers saw "clear explanation to be news" (Breed 1955, 264). Journalists ridiculed "the long report [...]: the 'think piece,'" calling the articles "thumb sucking" (ibid., 317), and continued "to revolve around the immediate, the flash, the shock, the punch" (ibid., 18). But event-centered, episodic news came under intellectual attack. French semiologist Roland

Barthes wrote in a 1964 essay that fact-based items—what the French call *Fait-Divers*—were not news, which occurs within a context in history, but mere filler, trivial facts made interesting by a suspicious or astonishing cause or by a peculiar coincidence and not requiring knowledge beyond the item itself (Barthes 1972; Boudana 2012). By the 1990s, *Fait-Divers* still seemed the U.S. model, buttressed by the idea that staying oriented is possible for "the body in a geo-astronomical space," in the words of semiologist Umberto Eco, because humans would "follow certain 'directions' [e]ven if the world were a labyrinth" (Eco 1990, 2). But U.S. journalism was changing *sub rosa,* in the background. Historian Robert Darnton, in a memoir of his time at the *New York Times* in the 1960s, recalled one reporter telling him, "The game is to sneak some color or interpretation past" copyeditors (Darnton 1990, 69). Over time the reporters succeeded.

Like a realist news story, the intellectual world as the twentieth century ended seemed to present two points of view. Anthropologist Clifford Geertz observed that the blurring of genres once separating fact from theory also blurred "what goes on in the world" (Geertz 2001, 179). From the liberal side, liberal news work pursues broad, complex purposes, follows rituals, performs like a medieval bard, and builds cultural memories and myths (Schudson 2003; Hartley 1982; Carey 1988, 1989; Meyrowitz 1991; and Corner 1995). Journalists abandoned the term *objectivity* in favor of terms like *fairness* and *balance,* which quickly became another interpretive quagmire. From the conservative side, social scientists wondered about "The Future of Fact" (Strange and Katz, 1998), but continued to push for a return to realism in the face of reform movements and critics (Fallows 1996). Whichever side journalists choose in the story of realism, the world around them had changed dramatically.

One way to choose is in everyday practice. For a century, journalists acted as professionals with access to hard-to-reach and authoritative information, writing analytical stories to explain events. If news professionals provide knowledgeable interpretations, the audience must be under-informed and requires others to judge what events mean (Carey 2002). The realist choice for news relies on top-down authority and a one-to-many style of public information. The resulting perception, that news lacks interpretation, blocks an understanding that sense making in American news accelerated. The contradiction between that imagined world and the growing thematic news coverage invokes another classic idea: that what societies define as real *is* real in its consequences (Thomas and Znaniecki 1927). The perception of an urgent need for journalists to make sense of a world too complex for average citizens to understand led inexorably to the expansion of explanatory news.

But emerging interactive media present news organizations with contradictory expectations. Instead of supplying what citizens cannot otherwise get, journalists seem like just another supplier among many competing on legacy media—cable outlets, talk radio, and comedy newscasts—and online (bloggers and aggregators). Instead of neutral observers in search of truth, journalists seem to play an on-stage role in manufacturing social knowledge (Goffman 1959). The new interactive world of news and its expectations are no better inherently. The realist perspective journalists prefer never quite reflected public intelligence or engagement with news media, and the new vision is also at odds with the conditions of public information: journalists working in legacy media continue to be primary suppliers of news, especially local political coverage. But the contradiction invokes another classic idea from sociology: that the social "definition of the situation," not some "nature" external to society, makes the world that people live into what it is (Thomas and Znaniecki 1927).

In the emerging interactive order, citizens become part of making a sensible world. A once-hidden collective social endeavor seems to emerge: the public informing itself. Journalists may come to play a smaller—but imaginably freer—role in the new media arena (Blumer 1969). News as an interaction, despite its still-shaky economic status, may liberate journalists to experiment with new forms drawn from local knowledge and available technologies. Competition may move from the news organization to the individual news worker, an uncomfortable and neoliberal economic transformation. But in the past, competitive pressures were always a one-size-fits-all argument, aligning news content expectations with the industry bottom line. The new world of news might provide journalists with other options for pursuing their professional aspirations.

Changing economic competition expresses something less apparent, a contest among ideas. The current crisis may strip journalists of the realism embedded in older definitions of news. Declining profitability for legacy news platforms may free the public from journalists as explainers. Both possibilities involve risk, but societies adjust rapidly. Changing news production and delivery techniques have continually unsettled the institutions entrusted with public information. If responsibility spreads across society, citizens may rise from being mere consumers and journalists may redefine themselves as independent news agents.

Works Cited

Adatto, Kiku (2008). *Picture Perfect: The Art and Artifice of Public Image Making.* Princeton: Princeton University Press.

Adkins-Covert, Tawnya et al. (2000). News in my Backyard: Media and Democracy in an "All American" City. *The Sociological Quarterly*, 41 (2), 227–244.

Armao, Rosemary (2000). The History of Investigative Reporting, In Marilyn Greenwald and Joseph Bernt (eds.). *The Big Chill: Investigative Reporting in the Current Media Environment*, 35–50. Ames: Iowa State University Press.

Bagdikian, Ben (2004). *The Media Monopoly.* Boston: Beacon Press.

Barnhurst, Kevin G. (2012a). "Trust me, I'm an innovative journalist," and Other Fictions. In Marcel Broersma and Chris Peters (eds.). *Rethinking Journalism: Trust and Participation in a Transforming Media Landscape*, 210–220. London: Routledge.

Barnhurst, Kevin G. (2012b). The Content of Online News in the Mainstream U.S. Press, 2001–2010. In Steve Jones (ed.). *Communication @ the Center*, 231–253. New York: Hampton Press.

Barnhurst, Kevin G. (2010a). Technology and the Changing Idea of News: 2001 U.S. Newspaper Content at the Maturity of Internet 1.0. *International Journal of Communication*, 4, 1082–1099, http://ijoc.org/ojs/index.php/ijoc/article/view/823/477.

Barnhurst, Kevin G. (2010b). The Form of Reports on U.S. Newspaper Internet Sites. An Update. *Journalism Studies*, 11 (4), 1–12.

Barnhurst, Kevin G. (2003). The Makers of Meaning: National Public Radio and the New Long Journalism, 1980–2000. *Political Communication*, 20 (1), 1–22.

Barnhurst, Kevin G. (2002). News Geography & Monopoly: The Form of Reports on U.S. Newspaper Internet Sites. *Journalism Studies*, 3 (4), 477–489.

Barnhurst, Kevin G. (1996). The Alternative Vision: Lewis Hine's Men at Work in the Dominant Culture. In Marsha Bryant (ed.). *Photo-Textualities: Reading Photographs and Literature*, 85–107. London: Associated University Presses.

Barnhurst, Kevin G. (1991). The Great American Newspaper. *The American Scholar*, 60, 106–112.

Barnhurst, Kevin G., and Diana C. Mutz (1997). American Journalism and the Decline of Event-Centered Reporting. *Journal of Communication*, 47 (4), 27–53.

Barnhurst, Kevin G., and Catherine A. Steele (1997). Image Bite News: The Coverage of Elections on U.S. Television, 1968–1992. *Harvard International Journal of Press/Politics*, 2 (1), 40–58.

Barthes, Roland (1972). *Critical Essays.* Evanston, Ill.: Northwestem University Press.

Blumer, Herbert (1969). *Symbolic Interactionism: Perspective and Method.* Berkeley: University of California Press.

Boudana, Sandrine (2012). The Unbearable Lightness of the Fait Divers: Investigating the Boundaries of a Journalistic Genre. *Critical Studies in Media Communication*, 29 (3), 202–219.

Breed, Warren (1955). *The Newspaperman, News, and Society*. Doctoral Dissertation, Columbia University.

Brinkley, Alan (2010). *The Publisher: Henry Luce and His American Century*. New York: Alfred A. Knopf.

Carey, James W. (1988). *Media, Myth, and Narratives*. Newbury Park, Calif.: Sage.

Carey, James W. (1989). A Cultural Approach to Communication. In *Communication as Culture: Essays on Media and Society*, 13–36. Boston: Unwin Hyman.

Carey, Jim (2002). What Does 'Good Work' in Journalism Look Like? *Nieman Reports*, 03/04/2014, http://www.nieman.harvard.edu.

Coblentz, Edmond D. (1954). *Newsmen Speak: Journalists on their Craft*. Berkeley: University of California Press.

Commission on Freedom of the Press (1947). *A Free and Responsible Press*. Report. Chicago: University of Chicago Press.

Connery, Thomas B. (1992). *A Sourcebook of American, Literary Journalism: Representative Writers in an Emerging Genre*. New York: Greenwood Press.

Corner, John (1995). *Television Form and Public Address*. London: Edward Arnold.

Corrigan, Dan (1999). *The Public Journalism Movement in America: Evangelists in the Newsroom*. New York: Praeger.

Cowley, Malcolm (1947). A Natural History of American Naturalism. *The Kenyon Review*, 9, 414–435.

Crossen, Cynthia (1994). *Tainted Truth: The Manipulation of Fact in America*. New York: Simon & Schuster.

Burgh, Hugo de (ed.) (2000). *Investigative Journalism: Context and Practice*. London: Routledge.

Darnton, Robert (1990). Journalism: All the News That Fits We Print. In *The Kiss of Lamourette: Reflections in Cultural History*, 60–93. New York: Norton.

Davis, Lennard J. (1983). *Factual Fictions: The Origins of the English Novel*. New York: Columbia University Press.

Dewey, John (1927). *Public and its Problems*. New York: Henry Holt & Co..

Doherty, Richard, and Kevin G. Barnhurst (2009). Controlling Nature: Weathercasts on Local U.S. Television News. *Journal of Broadcasting & Electronic Media*, 53 (2), 1–16.

Eco, Umberto (1990). *The Limits of Interpretation*. Bloomington: Indiana University Press.

Fallows, James M. (1996). *Breaking the News: How the Media Undermine American Democracy*. New York: Pantheon.

Fishkin, Shelley Fisher (1985). *From Fact to Fiction: Journalism and Imaginative Writing in America*. New York: Oxford University Press.

Fuller, Jack (2010). *What Is Happening to News?* Chicago: University of Chicago Press.

Fuller, Jack (1996). *News Values: Ideas for an Information Age*. Chicago: University of Chicago Press.

Geertz, Clifford (1980). Blurred Genres: The Refiguration of Social Thought. *The American Scholar*, 29 (2), 165–179.

Goffman, Ervin (1959). *The Presentation of Self in Everyday Life*. New York: Anchor Books.

Goldman, Peter et al. (1994). *Quest for the Presidency*. College Station: Texas A&M University Press.

Graber, Doris A. (1990). Seeing Is Remembering. *Journal of Communication*, 40 (3), 134–155.

Graber, Doris A. (2006). *Mass Media and American Politics*. 7th ed. Washington, D.C.: Congressional Quarterly Press.

Hallin, Daniel (1992). Sound Bite News: Television Coverage of Elections, 1968–1988. *Journal of Communication*, 42 (2), 5–24.

Harris, Roy J. (2007). *Pulitzer's Gold: Behind the Prize for Public Service Journalism*. Columbia: University of Missouri Press.

Hartley, John (1982). *Understanding News*. London: Routledge.

Haskell, Thomas (1998). *Objectivity Is Not Neutrality: Explanatory Schemes in History*. Baltimore: Johns Hopkins University Press.

Henry, William A., III. (1985). *Visions of America: How We Saw the 1984 Election*. Boston: Atlantic.

Houston, Brant (2010). The Future of Investigative Journalism. *Daedalus* 'The Future of News' Issue, 139 (2), 45–56.

Iyengar, Shanto (1991). *Is Anyone Responsible? How Television Frames Political Issues*. Chicago: University of Chicago Press.

Johnstone, John W. C., Edward J. Slawski, and William W. Bowman (1976). *The News People: A Sociological Portrait of American Journalists and Their Work*. Urbana: University of Illinois Press.

Kaplan, Amy (1988). *The Social Construction of American Realism*. Chicago: University of Chicago Press.

Kazin, Alfred (1942). *On Native Grounds: An Interpretation of Modern American Prose Literature*. New York: Harcourt Brace.

Kovach, Bill, and Tom Rosenstiel (2001). *The Elements of Journalism: What News People Should Know and the Public Should Expect*. New York: Crown.

Landers, James (2005). *The National Observer*, 1962–77: Interpretive Journalism Pioneer. *Journalism History*, 31 (1), 19–22.

Lang, Gladys Engel, and Kurt Lang (1983). *The Battle for Public Opinion: The President, the Press, and the Polls during Watergate*. New York: Columbia University Press.

Lee, Seow Ting, and Crispin C. Maslog (2005). War or Peace Journalism? Asian Newspaper Coverage of Conflicts. *Journal of Communication*, 55 (2), 311–329.

Lippmann, Walter (1922). *Public Opinion*. New York: Harcourt Brace.

MacDougall, Curtis D. (1938). *Interpretative Reporting*. New York: Macmillan.

MacDougall, Curtis D. (1972). *Interpretative Reporting*. 6th ed. New York: Macmillan.

Machin, David, and Sarah Niblock (2006). *News Production: Theory and Practice.* London: Routledge.

Marriott, Stephanie (2007). American Election Night and the Journalism of Assertion. *Journalism Theory, Practice & Criticism*, 8 (6), 698–717.

Marsden, George M. (2006). *Fundamentalism in American Culture.* New York: Oxford University Press.

McAuliffe, Kevin Michael (1978). *The Great American Newspaper: The Rise and Fall of the Village Voice.* New York: Charles Scribner's Sons.

McCarthy, Mary (1960). The Fact in Fiction. *Partisan Review*, 27 (3), 438–458.

McCourt, Tom (1999). *Conflicting Communication Interests in America: The Case of National Public Radio.* Westport, Conn.: Praeger.

Mears, W. R. (2009). Finding News Despite the Noise. *Journalism Studies*, 10 (3), 423–426.

Merchant, Carolyn (2004). *Reinventing Eden: The Fate of Nature in Western Culture.* New York: Routledge.

Meyer, Philip (1973). *Precision Journalism.* Bloomington: Indiana University Press.

Meyrowitz, Joshua (1991). The Questionable Reality of Media. In John Brockman (ed.). *Ways of Knowing: The Reality Club*, 141–160. New York: Prentice Hall.

Mosier, Nancy R., and Andrew Ahlgren (1981). Credibility of Precision Journalism. *Journalism Quarterly*, 58 (3), 375–381, 518.

Murphy, James E. (1974). *The New Journalist: A Critical Perspective* (*Journalism Monographs*, Vol. 34). Lexington, Ky.: Association for Education in Journalism.

Ogan, Christine L. (1982). Development Journalism/Communication: The Status of the Concept. *International Communication Gazette*, 29 (1/2), 3–13.

Orvell, Miles (1989). *The Real Thing: Imitation and Authenticity in American Culture, 1880–1940.* Chapel Hill: University of North Carolina Press.

Patterson, Thomas E. (2003). Diminishing Returns: A Comparison of the 1968 and 2000 Election Night Broadcasts. *KSG Working Paper Series No. RWP03–050*, 03/04/2014, http://www.hks.harvard.edu/presspol/publications/reports/election_night_report_2003.pdf.

Patterson, Thomas E. (2000). *Doing Well and Doing Good: How Soft News and Critical Journalism Are Shrinking the News Audience and Weakening Democracy—And What News Outlets Can Do About It.* Report published by the Shorenstein Center.

Perlmutter, David D. (2005). Photojournalism and Foreign Affairs. *Orbis*, 49 (1), 109–122.

Pizer, Donald (1998). The Late Nineteenth and Early Twentieth Centuries, 1874–1914. In *Documents of American Realism and Naturalism*, 3–17. Carbondale: Southern Illinois University Press.

Rosen, Jay (1999). *What Are Journalists For?* New Haven: Yale University Press.

Rosenfeld, Sophia (2011). *Common Sense: A Political History.* Cambridge, Mass.: Harvard University Press.

Rudner, Richard (1953). The Scientist Qua Scientist Makes Value Judgments. *Philosophy of Science*, 20 (1), 1–6.

Salgado, Susana, and Jesper Strömbäck (2012). Interpretive Journalism: A Review of Concepts, Operationalizations and Key Findings. *Journalism Theory, Practice & Criticism*, 13 (2), 144–161.

Schram, Martin (1987). *The Great American Video Game: Presidential Politics in the Television Age*. New York: Morrow.

Schudson, Michael (1992). Water*gate in American Memory: How We Remember, Forget, and Reconstruct the Past*. New York: Basic Books.

Schudson, Michael (2003). *The Sociology of News*. New York: Norton.

Seib, Philip M. (2007). *Broadcasts from the Blitz: How Edward R. Murrow Helped Lead America into War*. Washington, D.C.: Potomac.

Shafer, Richard (1991). *Journalists for Change: Development Communication for a Free Press*. Manila: Philippine Press Institute.

Siebert, Frederick S., Theodore Peterson, and Wilbur L. Schramm (1956). *Four Theories of the Press*. Urbana: University of Illinois Press.

Sontag, Susan (1977). *On Photography*. New York: Farrar, Straus & Giroux.

Stauber, John, and Sheldon Rampton (2002). *Trust Us, We're Experts! How Industry Manipulates Science and Gambles with your Future*. New York: Tarcher Putnam.

Steele, Catherine A., and Kevin G. Barnhurst (1996). The Journalism of Opinion: Network Coverage in U.S. Presidential Campaigns, 1968–1988. *Critical Studies in Mass Communication*, 13 (3), 187–209.

Strange, Jeffrey J., and Elihu Katz (eds.) (1998). 'The Future of Fact' Issue. *Annals of the American Academy of Political and Social Science*, 560 (1).

Swinford, Dean (2001). Defining Irrealism: Scientific Development and Allegorical Possibility. *Journal of the Fantastic in the Arts*, 12 (1), 77–89.

Thomas, William I., and Florian Znaniecki (1927). *The Polish Peasant in Europe and America*, New York: Knopf.

Trachtenberg, Alan (1977). Ever—the Human Document. In *America and Lewis Hine: Photographs 1904–1940*, 118–138. New York: Aperture.

Trow, George W. S. (1999). *My Pilgrim's Progress Media Studies, 1950–1998*. New York: Pantheon.

Väliverronen, Jari, and Risto Kunelius (2012). From Reporting to Storytelling: Mediating the Political Public Sphere in Finland From the 1960s to 2005. In Steve Jones (ed.). *Communication @ the Center*, 215–229. New York: Hampton Press.

Weaver, David H., and G. Cleveland Wilhoit (1991). *The American Journalist: A Portrait of U.S. News People and Their Work*. Bloomington: Indiana University Press.

Weaver, Paul H. (1994). *News and the Culture of Lying*. New York: Free Press.

Wicker, Tom (1978). *On Press*. New York: Viking.

Wilner, Isaiah (2006). *The Man Time Forgot: A Tale of Genius, Betrayal, and the Creation of* Time *Magazine*. New York: Harper-Collins.

Winter, Ella, and Herbert Shapiro (eds.) (1962). *The World of Lincoln Steffens*. New York: Hill and Wang.

Wolseley, Roland E., and Laurence R. Campbell (1949). *Exploring Journalism: With Special Emphasis on its Social and Vocational Aspects.* New York: Prentice-Hall.

Contemporary or Commercial? The Impact of Innovative Image Reproduction Techniques on the Imagery of Nineteenth Century Mass Media in Germany

Rita Gudermann

Introduction

Long before periodicals and books were lavishly illustrated with photographs, a wide range of other innovative illustration methods developed during the nineteenth century. Many of the new techniques were short-lived, but others led to impressive illustrated works offering a completely new visual approach to everyday life. Though being initially complex and costly, they opened up new worlds of visual content. New printing businesses were launched and with them a specialized supply industry. New technical literature had to be written and distributed, new subjects had to be taught at technical schools and at publishing houses. This development took place in all major industrial countries and yet has been undeservedly overlooked by researchers.

This paper is part of my research project "Content as a Commodity" (Original title: "Ware Information"), in which I describe the process of commercialization of content as a virtual product, from the 1830s to 1945. In the following, I will focus on the role of new illustration techniques in shaping the publishing industry in Germany from the 1830s to the 1940s, which is from the first appearance of the penny magazines to the dawn of the TV age. I will argue that the adoption of new illustration techniques was part of a syndication strategy that enabled the launch of new companies and contributed to the economic well-being of others. I will start by giving a short overview of different illustration techniques and key technological innovations. To show the impact they had on publishers' strategies, I will present two case studies of Leipzig-based publishing houses. Furthermore, I will address the question of when this process started and which techniques had been influential. What was the impact of the new technologies on publishers and which of their strategies proved to be successful?

Time Line: Innovative Illustration Techniques and their Impact on the Printing Industry

The spectrum of nineteenth and early twentieth century illustration techniques was impressive. It contained aquatints, autotypes, chromolithographs, collotypes, copper engravings, copper intaglios, lithographs, paper embossings, photogravures, stereotypes, wood cuts, wood engravings and others (Krüger 1929). Today, many of these techniques are completely forgotten. They have been gradually replaced by the ubiquitous photography that presented prints in fine grained half-tone and dominates the media until today.

The astonishing range of different technical terms reflects flourishing new industries and skills: Xylographic and steel engraving ateliers, galvanization workshops, and several specialized supplying industries (e.g. Köhler 1903). As the printing industry globalized in the nineteenth century, most techniques were well-known in industrial countries, and leading printer shops adopted them to expand their business activities.

We can distinguish four different types of techniques used for book and magazine illustrations: 1. Letter press printing methods (*Hochdruckverfahren*) such as wood cuts and wood engravings, 2. Gravure printing methods (*Tiefdruckverfahren*) such as copper and steel engravings and etchings, 3. Planographic printing methods (*Flachdruckverfahren*) such as lithographs and chromolithographs, and 4. Photographic printing techniques (*photographische Verfahren*) such as autotypes. In the following I will focus on the economically most important procedures and arrange them on a time line.

Wood cuts first flourished in the fifteenth century, when flysheets were the mass media of the illiterates during the peasant uprisings and the protestant reformation. Made from wooden printing plates, these pictures could be printed on the letter press. This production method was easy and cheap, since individual images could be added to the type tray and printed together with the letters in one step. Though these pictures were often colored, they remained rough and raw with no space for detail and refinement. There was an enormous difference between popular books and bibliophilic incunables of the fifteenth century, the latter illustrated with sometimes lovingly manually colored copper engravings.

In the sixteenth century, techniques improved slightly, due to the increased use of boxwood for the printing plates. It allowed finer engraved lines and more detailed depictions. A new type of book emerged: The picture book that was not meant for children but for a semi-literate adult public. The images became more realistic. They contained foundations to depict light and shadow and thus competed with copper engravings. Therefore, subsequent manually colored wood cuts became less common. Most books of this time contained illustrations, in particular decorative capital letters (Janzin and Güntner 2007).

The bloom of *copper engravings* started in the seventeenth century and soon outweighed the importance of wood cuts. The technique had been known since 1440, i.e. as far back as the wood cut. But since gravure printing methods made it necessary to print text and images intricately, in two stages of work, they were never able to replace the wood cut completely and remained the illustration method for the rich and well-educated (Janzin and Güntner 2007, 257–259).

Books illustrated with copper engravings had their heyday in France in the eighteenth century and spread from there to other European countries. For publishers, the time-consuming and costly production of books with copper engravings was a risk, because plagiarists were often faster in bringing illustrated works with less elaborate engravings to the market. Since no satisfactory methods were found to speed up the process, copper engravings never became the main illustration method for mass production.

The *steel engraving* technique, developed by the American mechanical engineer and inventor Jacob Perkins (1766–1849), was another gravure printing method, based on extremely durable steel plates that could be used for several thousand copies. The technique was time-consuming and mainly used for the production of stamps and banknotes, but also for book illustrations (see Fig. 1). Another advantage was the option of reproducing the original printing plate. After the engraved original had been hardened, it could be copied by transferring the engraver's lines onto another steel plate which was hardened after duplication and was almost indistinguishable from the original. Therefore this technique was mainly used for the reproduction of book illustrations in high circulation production such as encyclopedias (Krüger 1929, 216–219; Van der Linden 1990, 115–116).

During the eighteenth century, when gravure printing had its heyday, wood cuts were only used for vignettes, borders, and capital letters, and as rough illustrations in common literature and calendars—the reason for their bad reputation among the educated public.

Fig. 1: Steel engravings were used for prints with high circulation—the example shows a postcard depicting Napoleon
(Source: Postcard, not dated [c. 1900], Collection Quagga Illustrations)

Fig. 2: Example of a nineteenth century wood cut image from the Pfennig-Magazin depicting the extinct Mauritian bird Dodo (Source: The Penny Magazine [1833], Cover of Issue 27)

Yet in the nineteenth century, wood cuts had a temporary comeback with the rise of the first illustrated mass production, the penny magazines in the 1830s (Fig. 2). To shorten the time needed to produce a single wood cut, several xylographers worked on different parts of the same picture, which were glued together afterwards. With the help of metal stereotypes (known as clichés), cast from a papier-mâché mold, whole magazine layouts could easily be reproduced. These techniques were an important factor in making the penny magazines affordable for mass production (Gebhardt 1989).

During the nineteenth century, *wood engravings*, a crucial improvement of the wood cut, became the most important technique of printing illustrations, since their quality didn't fall short of that of the copper engraving (Fig. 3). They flourished from the 1840s to the end of the century, when they were almost completely replaced by autotypes.

Fig. 3: The extinct bird Dodo reproduced as wood engraving, published in an illustrated encyclopedia (Source: Spamers Illustrirtes Konversations-Lexikon [1893]. Nachschlagebuch für den täglichen Gebrauch. Wohlfeile Ausgabe, Vol. 3, 1215. 2nd rev. ed. Leipzig: Spamer)

Like wood cuts, wood engravings were made of strong and fine grained boxwood, but the printing block was handled by the engraver instead of the knife. Besides, the engraver used cross-grained instead of side-grained wood. Since full-grown specimens of box trees were rare, wood engravings had to be small, until engravers found a way of combining several small printing plates to make larger pictures. The blocks were mainly used to make black-and-white prints, combining the fine-grained tonality of copper printing methods with the convenience of wood cuts. Wood engravings had the advantage that they could be printed together with the text on the letter press in one operation. Though this method was developed by several printers at the same time, the credit for the invention is usually attributed to the Briton Thomas Bewick (Hanebutt-Benz 1984, 607–611; Uglow 2006).

The production of wood engravings was time-consuming: A picture reaching over two columns of the *Illustrirte Zeitung* consisted of 20 blocks of 10 square centimeters each. This involved 60 working days for the xylographer. However, the demand for wood engravings was huge, leading to further innovations to make the method faster and cheaper: Since fine lines formed the tonality of wood engravings, cutting machines were developed which cut large sections of, for instance, equally toned parts of sky or water in a fragment of the time an engraver would have needed (Weise 2003, 73).

To preserve the precious and expensive boxwood blocks, stencils were invented that could be used to reproduce the original. Those stencils were produced by galvano-technique, an electro-chemical method that produced an accurate copy of the original, which could subsequently be used instead. To the inexperienced eye, it was almost impossible to distinguish whether the print was made using the original boxwood block or the galvano method. New manufacturing companies emerged who specialized in reproducing boxwood blocks. Others focused on the supply of electro-chemical materials and machines.

The impact of this innovation cannot be overestimated: A new and inexpensive way was found to produce the detailed black-and-white prints that became the hallmark of popular magazines and books in the second half of the nineteenth century. They made it possible for illustrations to become an important part of the emerging mass production sector (Schulze 1926, 10). During the time that followed, family magazines with their specific mixture of texts and images developed. Pictures were used to increase interest and to provide a complete impression of the chosen subject. Together with the page layout, they produced the typical appearance of the family magazine and constituted its character (Duttenhöfer 2013, 309–311).

Contemporaries were well aware of this development: They described wood engravings as a new, democratic method to meet the professional's needs, as well as the children's wishes (Hecht 1897, 56). Editors considered themselves as playing an important part in the distribution of art, knowledge, freedom, and humanity. Yet there were also critics' scathing remarks about a new tendency towards superficial reading habits and casual use: "How comfortably the image offers the half asleep, the indifferently scrolling, the lethargic and oversaturated reader superficial pleasures, without the bothersome effort of reading! And this alone is what a great part of our journalistic illustrations aim for" as a German observer of the media market wrote (Oldenberg 1859; quoted from Haug and Kruse 2004, 9).

It was the wood engraving that paved the way for the use of photographs as magazine and book illustrations. In order to print photos, they first had to be remade as xylographs. From 1860, photographs could be used directly for the preparation of the wood block by using a light-sensitive emulsion (collodion technique). For the first time, the Britons Vincent Beechy und Robert Langton (1854) and the Italian Francesco Ratti (1856) succeeded in transferring a photograph directly onto the boxwood printing plate. In 1859, the Briton William Spence was granted a patent for this method. He prepared the wood blocks and soaked them in a silver dilution. In 1861 the *Illustrated London News* presented the first wood engraving from a wood photograph showing the Cathedral of Exeter, made by the xylographer T. Bolton. The new method allowed further attention to detail and made the reproduction of paintings attractive. Images proudly bore the evidence 'made from a photograph.' In Germany the first photo-xylograph appeared in issue 20 of the popular magazine *Gartenlaube* in 1865. By the mid-1870s the photo-xylography had become an established illustration method which was a huge step towards authenticity. However, the making of magazines was still a tedious process. At the end of the 1870s, it took three weeks from the first concept and design to the finally printed issue, despite the invention of the high pressure rotation printing machine in 1869 in Germany. It was not until 1879 that a rotation printing press, which was built especially for the magazine *Illustrierte Welt,* met all requirements of the image printer (Weise 2003, 71–73).

Though wood engravings left almost nothing to be desired in terms of accuracy, they lacked an essential aspect: color. Certainly, expensive manually colored books existed, but they never became mass produced. Driven by the ambition to invent forgery-proof banknotes, the Briton William Congreve found a way in 1822 of printing several colors on top of each other, permitting a variety of colors in one single wood engraving image. But the technique was expensive and did not make its way into the repertoire of the average printer (Hanebutt-Benz 1984, 611–613).

To add a spark of color to popular books and magazines, it took a well-established printing technique, the *lithograph,* and adapt it to the requirements of color printing. Made out of Solnhofen limestone, printing plates were extremely robust and could be used for several editions. For *chromolithographs,* (Fig. 4) up to 20 colors were applied onto the stone in a complex procedure, each of them coloring only a few parts of the picture that had to be imprinted and dried several times to complete it (Krüger 1929, 260–291).

Fig. 4: Chromolithographs could also contain only two colors allowing a fine tonality: Reproduction of Albrecht Dürer's Crucifixion. (Source: Brockhaus' Konversations-Lexikon [1898], Vol. 5, 628–629. 14th rev. and enh. ed. Leipzig, Berlin, Wien: F. A. Brockhaus)

This process had already been described by its inventor Alois Senefelder, but it was costly and could not be integrated into the process of printing the text of the book or magazine. The colored images had to be inserted manually into the book block and remained a foreign body with their thick and oily appearance and texture. Nevertheless, this method was common from around 1880 to 1910 for enhancing encyclopedias, children's and popular non-fiction books. Chromolithographs made color illustrations affordable for the masses. And the public honored the printer's effort not only by buying books that contained 'chromos' with shiny colors, but also by indulging in colorful stationery and paper models and by collecting scraps, business cards, or postcards.

It took another technological step to integrate photographs into magazines and books. The *autotype*, developed by the German copper engraver Georg Meisenbach from Nuremberg, transferred the toning of black-and-white photographs into etchable and printable dots. Meisenbach was granted a patent in May 1882, and the first autotype was published in the *Zeitschrift des Kunst-Gewerbe-Vereins in München* in October of the same year. In the beginning, this method was mainly used for the reproduction of artistic objects. This was paradigmatic: Photographs were not yet seen as a means of greater vogue, but of greater detail and accuracy of static objects. Only towards the end of the century, when exposure times of 1/50 or 1/25 seconds were possible, did the snapshot enter the picture and change the face of media forever. Since autotypes were remarkably fast to produce, they soon replaced wood engravings. In 1903 a business magazine for printers stated in expert terms: *"In our nervous time we no longer appreciate the slow and thorough skills of the engraver" (Deutscher Buch- und Steindrucker 1903/04, 833;* quoted from Weise 2003, 100).

Autotypes were less labor-intensive and with six *Pfennig* per square centimeter in 1903 already significantly cheaper than xylographs (ten *Pfennig* per square centimeter). Thus autotypes dominated mass media in the 1890s (see Fig. 5). Magazine and newspaper producers embraced the new technique and their imagery was soon dominated by photographs. New magazines such as the *Berliner Illustrirte Zeitung* (established 1892) and the magazine *Vom Fels zum Meer* (established 1894) abstained from wood engravings from the beginning and used autotypes exclusively. As a result, a new visual language emerged. Photo journalism arrived, by means of photography, with its claim to report newsworthy events in politics, culture and society to shed light on the climax of an event, as well as its background, and to add an emotional

Fig. 5: The enlargement of the picture shows the raster that is used for the reproduction of a photograph in print (Source: Krüger 1929, 94–95)

aspect. This development, which was crucial in the twentieth century media, has been described by Jens Jäger and others (Jäger 2000, 104–112; Weise 1989a, 1989b, 1990, 1994, 1996).

After publishers' and printers' workshops had produced the majority of popular pictures initially, the rise of autotypes created a new market for images. With the beginning of press photography in the 1880s, the media world saw this market rapidly increasing. Freelance photographers sold their products directly to newspapers and news agencies. New independent picture libraries (such as Zander & Labisch Illustrations-Photographen in Berlin) found their niche in the market. At the beginning of the twentieth century separate picture archives (whose history has yet to be written) were established as part of publishing houses (such as Ullstein, from 1900 onwards) and news agencies such as Reuters or the BBC Picture Archive established in 1922. Later, intermediaries played an important role within the distribution of pictures and in 1920 the first independent picture agencies came into being (see list of agencies in Dietze 1933, 258–262). The US-American agency Keystone opened its doors in Germany in the 1920s (Weise 1999; Bruhn 2007; Caraffa 2011; Zierenberg 2013).

The Impact of New Illustration Techniques on Publishing Houses: Two Debuts in Leipzig

To throw some light on the impact of new illustration techniques on publishers' strategies, I will refer to the foundation of the publishing houses of J.J. Weber and of Otto Spamer in Leipzig. Despite their success, neither of the two companies survived the changes caused by the chaos of the Second World War and the postwar period. Today they are almost forgotten. Though their bookkeeping files are only fragmentary, their rich portfolios provide valuable information on the strategies of trade publishers. To reconstruct the programs of the two publishing houses, I compiled several databases using original sources such as the publishers' own catalogues, brochures and *Festschriften* (collection of articles) as well as library databases and catalogues, primarily from the *Deutsche Nationalbibliothek* (DNB), the *Zeitschriftendatenbank der Staatsbibliothek zu Berlin* (ZDB), the GBV (*Gemeinsamer Bibliotheksverbund*, and the *Stiftung Preußischer Kulturbesitz*), the *Zentrales Verzeichnis antiquarischer Bücher* (ZVAB), the *British Library Integrated Catalogue and the Bibliothèque nationale de France* (BnF). This approach provided an impressive database of the publishers' portfolios. The collected data contains several thousand publications, each with year, title, author, translator, series, number of pages and illustrations, price, etc. The aim was to reconstruct the output of the publishing houses as completely as possible and to present statistics of their production, stating pages and images.

The publishing house J.J. Weber in Leipzig existed from 1834 until 1947. It was founded by Johann Jakob Weber (1803–1880), a former bookseller from Basel, who came to Leipzig where he was entrusted by the well-known French bookseller and publisher Martin Bossange with the establishment of the German version of the penny magazine which had already been successful in Britain and France. During several years in Paris, Weber made a number of international business contacts and met sponsors. He based his own enterprise on foreign licenses, especially the works of Charles Dickens. Having been part of the vanguard of the German *Pfennig-Magzin*, Weber was well aware of the significance of new illustration techniques and technological innovations. Thus he imported the new black and white letter press printing technique, i.e. wood engravings from France and England, and successfully launched a program of illustrated works that consisted of fiction and political literature, popular reference books in small formats, tourist guides and journals. During the early years, he had to import wood blocks, engraving materials and even the paper that was best suited to print illustrations. He then published the first comprehensive work illustrated with wood engravings in Germany, Franz Kugler's highly praised *Geschichte Friedrichs des Großen*, with illustrations of the artist Adolf Menzel. Since it proved to be difficult to get as many xylographs from England and France as needed, he established his own xylographic workshop and entrusted the congenial Eduard Kretzschmar to lead this department. The xylographic atelier became the focal point of the company and implemented numerous innovations. Many xylographers were trained at his workshop (Weber 2003).

Being the employer of his editors and xylographers, Weber had almost complete control of the contents of his productions. A cornerstone of Weber's success was the launch of the *Illustrirte Zeitung*, a popular weekly family magazine. It existed for more than a century from 1843 to 1944 and was the first magazine in Germany illustrated by wood engravings. With the *Illustrirte Zeitung,* Weber followed the example of the *Illustrated London News* (1842) and the French magazine *L'illustration* (1843). In the first issue of the magazine, Weber proclaimed a new fact: "Whatever happens in the entire known world […] if it is of general interest we will offer our readers weekly reports" (*Illustrirte Zeitung* 1843 [1], 1–2).

The *Illustrirte Zeitung* was again a forerunner when it came to adopting the innovation of the autotype. The first autotype was published in the *Illustrirte Zeitung* on October 13, 1883; in 1884 the first snapshots of a military manoeuver by Wilhelm II by Ottomar Anschütz were shown. From 1883 to 1890 the *Illustrirte Zeitung* published 206 autotypes, from 1891 to 1900 by this time 1800 photographs, and between 1900 and 1906 more than 5,000. At this time the *Illustrirte Zeitung* printed a total of ca. 1200 illustrations per year. In 1900 photographs for the first time outnumbered drawings. In 1905 the publishing house established a 'chemigraphical department' for the production of autotypes. The xylographer atelier, once the heart of the company, was closed down in 1906 (Weise 2003, 99–100).

Despite Weber's impressive performance on the media market, timing was a real problem. Technical restrictions led to extremely long periods of time between the actual event and its depiction in a magazine. A fire in the Berlin opera house on August 18, 1843 demonstrates this. Two weeks after the incident, on September 2, a picture of the opera house before the fire was published in the *Illustrirte Zeitung*. Two weeks later, on September 16, the magazine contained a picture showing the opera house in flames, and a whole month later, on October 14, an engraving of the ruins was revealed.

Initially the events were often reconstructed, based on eye-witness accounts. In 1850 a network of on-site corresponding illustrators appeared, as well as illustrators who covered special aspects such as wars, natural disasters, political or social subjects. The *Illustrirte Zeitung* asked their readers repeatedly to send in drawings of the latest events to improve the reality of its news coverage (Weise 2003, 70–71).

During the time of its existence, the publishing house J. J. Weber had an output of an average of 40 books and 80 magazines/journals annually, which gained Weber the reputation of an honorable old-school publisher. The founder's heirs followed his path with illustrated popular fiction and non-fiction books, journals and art reproductions, until the company was destroyed in an air raid in 1944 and officially concluded its activities in 1947.

The publishing house of Otto Spamer in Leipzig was founded in 1847, 14 years after Weber's, and lasted until 1948. The founder Otto Spamer (1820–1886) was a former bookseller and employee at J. J. Weber. Though he was supported by his former employer and mentor, he had a very difficult start. He was ten years too late: The market for wood engravings was saturated. All major publishers had adopted this technique, and Spamer had to sell odd goods like insecticide before his printed products took off. Another experiment of survival in a well-established market was non-fiction books which Otto Spamer wrote himself under the pseudonym Franz Otto. His success story began when he found his niche with color illustrations: He used chromolithographs in encyclopedias and books for young people. This was when he found out that vivid covers, wood engravings and chromolithographs made his books sell, especially when wrapped in sophisticated and colorful frontages. In 1868 he expanded his business by adding a bookbindery and in 1877 his own printer shop. He included illustrated encyclopedias, books for adolescents, technical books and journals in his program, and produced an output averaging ten books and five journals annually. Though his path proved to be harder than Weber's, he was eventually no less successful then his mentor. His successors perpetuated his achievements by publishing illustrated books and journals until the company was closed down in 1943 (Schmidt 1908).

Syndication vs. State-of-the-Art Strategy

What is the pattern in the histories of the publishing houses J. J. Weber and Otto Spamer? Did they follow similar strategies and which of these proved to be successful?

First of all, it is obvious that both enterprises were success stories, based on the production of richly illustrated books and journals, and high quality illustrations for the masses. Both success stories can be considered technology motivated, i.e. based on the adaptation of some of the new illustration methods. Whereas Weber focused on wood engravings and later on autotypes, Spamer chose chromolithographs and colorful covers to make his collection stand out. Weber was also successful in adapting a magazine format to the German market, which had proved to be popular in other European countries, and skillfully purchased foreign book licenses. A formula for success for both parties was the strategy of vertical integration, i.e. the establishment of xylographic, lithographic and photographic workshops and book binderies within the publisher's own company.

Less obvious is that both publishers were successful in the syndication of texts and images: In the course of their long existence they built a pool of multi-faceted content (especially engravings, but also novels, riddles and other kinds of texts) that could be used with different products (magazines, encyclopedias, books, calendars, brochures, etc.). Illustrated chronicles, almanacs, calendars and special editions too can be seen as subsequent use of already published material. And there were many of them, take, for example at Weber's publishing house *Illustrirte Chronik*, 1848–1854, *Illustrirter Kalender*, 1846–1881, *Illustrirtes Weihnachts-Album*, 1868–1874 (Christmas editons) and *Illustrierte Weltkriegschronik*, 1914–1920 (war chronicles), *Novellen-Zeitung*, 1844–1873 (collections of novels) and *Meisterwerke der Holzschneidekunst*, 1879–1904 (art reproductions). Another strategy was to publish popular technical literature that could be updated and enhanced by increased and improved texts and images with every new edition, thus staying 'hot-off-the-press' and worthwhile to the public. Many of their titles reached more than ten editions, which demonstrates that the publishers found a profitable way to re-use content (since a title proves to be more profitable the more editions it attains). However neither of the two publishers seems to have been successful with selling book licenses due to a lack of copyright control (Spamer 1872, 97).

Very often the richly colored books were accompanied by plain and artless versions of the same work. Instead of printing novels of Charles Dickens or Daniel Defoe only as a plain text version, for instance, the publishers could offer the same book enhanced with black-and-white illustrations, decorated with some chromolithographs or presented as a richly decorated and lavishly illustrated leather-bound hardback, combining colored and black-and-white pictures, maps, vignettes, initials etc. This was part of a differentiation strategy, aiming at different readers. Thus new illustration methods found their place in the publishers' syndication strategy.

Yet the two publishers were also successful in syndicating their content to other entrepreneurs. This started in 1846, when Weber offered the engravings already published in the *Illustrirte Zeitung* for syndication (Verzeichniss 1846). In 1858 Spamer followed with an illustration and cliché catalogue (Spamer 1858). In 1896 the catalogue *Bildermarkt der Illustrirten Zeitung* appeared, offering pictures of the events of the week in order to sell them with captions. This development—which also applied to other publishing houses—was the beginning of a flourishing market of picture licenses dating from the middle of the century, when images (clichés, printing plates) were discovered as a marketable product. At the same time, authors such as Theodor Fontane, Karl May and Hedwig Courths-Mahler, to name but a few, realized that they could increase their earnings by selling their products in installments, that is as serialized novels in several magazines and later as hardbacks (Graf 2003, 81–92). Thus publishers learned that syndication was a means of survival and prosperity without relying on capricious bestsellers and unpopular mixed computations, i.e. the necessity to finance works of art with trashy novels and shallow but popular non-fiction books (Bergk 1983, 13–14).

How did this strategy fit into emerging journalism? With the rise of photography, new journalistic formats developed, like news coverage with its claim to up-to-dateness, immediacy and individuality as characteristics. How did the rise of professional journalism affect the publishers' syndication strategies? In the following I will analyze the relation of standard images and 'latest' images, published in the *Illustrirte Zeitung*. I define 'stockable' images as those that are suited for multiple use, whereas 'news' or 'event' images will presumably be published only once. Of course, both categories overlap sometimes and a thoroughly conducted analysis of the flow of pictures that were positively recycled will have to be conducted to confirm my findings.

For this purpose, random specimens of the *Illustrirte Zeitung* were analyzed.[1] The images were categorized by their content: Specimens were categorized as 'news' or 'event' pictures if they illustrated newsworthy occurrences like assemblies, military action, battles, catastrophes, theatrical productions and so on. Portraits, buildings/interiors, fashion, landscapes, emblems/medals, cartoons, vignettes, technical illustrations etc. were categorized as 'stock' images. A short overview of the findings is presented in the following table.

	1/1843	*57/1871 (Jul.–Dec.)*	*94/1890*	*142/ 1914*	*173/1930 (no. 4425–4437)*	*1943 (no. 4966–4978)*
Σ images	693	404	754	907	894	1086
Σ event	104	89	92	626	281	483
Σ stock	589	315	662	281	613	603
% event	15.0	22.0	12.2	69.0	31.4	44.5
% stock	85.0	78.0	87.8	31.0	68.6	55.5

Tab. 1: Relation of news and stock pictures in specimens of the Illustrirte Zeitung, 1843–1943

What is obvious is that the magazine used stock images where possible. Only during the First World War specimens does the percentage of event images exceed the percentage of stock images. To enhance the up-to-dateness of the artwork during both wars, the editors made broad use of illustrations (not photographs) of events. Though these images depict violently vivid war scenes like battlefields, sinking warships or burning houses, this may have served to soften the effect of the sight of dead and wounded soldiers, too.

The number of news images greatly increased due to these illustrations, but stock photographs still had to serve as substitutes. Following the riots against the German embassy in St. Petersburg a simple photograph of the building's front was shown (*Illustrirte Zeitung* 1914 [3711], 305).

It can be said that the editors of the *Illustrirte Zeitung* tried to avoid up-to-dateness where possible. Although they generally claimed to cover up-to-date topics in their weekly magazine, the pictures they used remained 'stan-

1 *Illustrirte Zeitung*, 1/1843, 57/1871 (Jul.–Dec.), 94/1890, 1/1914, 173/1930 (No. 4425–4437), 1943 (No. 4966–4978).

dardized,' i.e. characterized by the option of producing them in advance and by their renewability in other media.

The *Illustrirte Zeitung* published portraits of celebrities, such as members of the royal family or other aristocratic houses, well known politicians, authors or scientists, all of them a staple of pictures. Another group of frequently used images were views of monuments, buildings, interiors, cities and landscapes. City maps and ground plans or genealogical tables were added. Illustrations depicting movement were often of monumental character and showed mainly historical scenes like ceremonies, military action, inaugurations, marriages, deathbeds or funerals. The *Illustrirte Zeitung* printed also encyclopedic (and distinctively static) pictures of plants, animals, physical subjects, techniques, fashion, furniture and character images (family scenes etc.). All these illustrations stood, in a way, for an idea, such as the 'concept' of piano playing (1843 [1], 124) or the 'concept' of ironing (1861 [941], 38).

These illustrations were never meant to show a specific moment, which is still true of today's pictures: The media has an insatiable hunger for generic images. For most of the nineteenth century, the pictures of the *Illustrirte Zeitung* were not even supposed to show a specific person. The faces of people in these generic pictures are flat and unrecognizable. This changed abruptly towards the end of the century when autotypes depicted the individuality even of people in the background. And yet, except for very rare documentary features, the pictures in family magazine claimed to show the truth not through individuality and specificity, but through significance and generality. Moreover, they gained legitimacy by the sheer fact that they were printed in a magazine and therefore had to be true, since the printing of images was a complex and expensive procedure, a fact the public was well aware of. Since the 1880s (i.e. before the copious use of photographs) the images show a greater liveliness and immediacy, be it a wood engraving of a beauty from the Balkans or a festive ball. What is missing among these pictures until the emergence of the autotype is the snapshot, the individual view of the contemporary witness, the life depiction, i.e. those captions that are the foundations of today's photo-journalism. Thus the publisher's new enthusiasm for standardization clearly clashed with another trend: The flourishing of journalism, i.e. the modern industry of news coverage, whose first ambition is to be as up-to-date as possible. The expensive new illustration techniques had to recuperate costs incurred, not only by high circulation figures of the illustrated media, but also by diverse use. This phenomenon changed only gradually towards the end of the century with the appearance of autotypes.

Subsequently, pictures had to be produced with an eye to their generality, they had to fit into a pool of universally usable content. As a result, illustrations that made their way through the media landscape often developed into icons. An example is the Dodo, a bird from the island of Mauritius, which became extinct towards the end of the seventeenth century (see Fig. 2 and 3 above). Presented in profile, the image of the bird (which none of the illustrators had ever seen) was copied so often that it claimed a truth that in fact never existed. This changed when autotypes became ubiquitous and showed people, animals or subjects in motion with distinguishable features. Yet in the course of this process, the single image lost its iconic character and became a commodity, forcing its producers to deliver more and more of the mundane product.

Conclusion

The in-depth analysis of the success stories of two Leipzig-based publishing houses has shown that publishers found effective ways to survive and thrive on the media market by adapting technological innovations and by being the first with high quality illustrations. But innovative new illustration techniques were not enough: Publishers had to find a formerly neglected demand from the public or, generally speaking, to create a new audience (as Weber did with middle class magazine readers and Spamer with adolescent readers) and a way to provide a cost-effective means of satisfying this demand by mass production. Thus, several illustration techniques, though technically promising, fell by the wayside because they were too expensive or too slow. Others, such as chromolithographs, though costly and time-consuming as well, offered the benefits of mass reproduction and were therefore adapted. Overall, new methods of printing illustrations provided publishers with the means to satisfy public demand for lavishly illustrated and richly decorated books and periodicals.

Since a detailed reconstruction of the flow of images through a variety of different media would be a research project on its own, several questions remain. Firstly, my findings have to be confirmed by further data analysis and international comparisons. It looks promising to track down individual items of content (text, images) through a variety of media products, media enterprises, and countries. Another desideratum is an in-depth qualifying analysis which explores publishers' costs and profits.

However, several findings can be stated: During the 19th century especially, those publishers profited who already owned a rich pool of content assets which could be syndicated. To flourish under these new conditions, in the first place publishers had to understand their content as assets and then sell it in a piecemeal form as often as possible. By syndicating text and images they laid the foundation of the flourishing content business of today. This phenomenon has not changed until today: Magazines consist largely of stock material illustrating an event (for example celebrities or a seasonal event). The new industry of image libraries satisfies this hunger for stock. Not only today but also in the pre-photography period, standardization was more promising as a strategy than up-to-dateness. The niche for actuality opened up with the rise of documentary photography and daily journalism. Being able to capture the specific moment became the achievement of the professional photo-journalist. The role of the publisher changed, too: Had he been the master of the written and visual content of his medium before, he had to face new players on the media market: freelance authors, photographers and their agencies. If it wasn't for the unique shot, visual content had become widely available, but also more casual and fugitive, less iconic and less re-usable.

Works cited

Bergk, Johann Adam (1983 [1825]). *Der Buchhändler oder Anweisung, wie man durch den Buchhandel zu Ansehen und Vermögen kommen kann*. Heidelberg: Winter.

Bruhn, Matthias (2007). Tarife für das Sichtbare. Eine kurze Geschichte der Bildagenturen. *Fotogeschichte*, 105, 12–25.

Caraffa, Costanza (ed.) (2011). *Photo Archives and the Photographic Memory of Art History*. Berlin: Deutscher Kunstverlag.

Dietze, Carl (1933). *Presse-Illustrations-Photographie. Fachweiser für die Verwertung der Gebrauchsphotographie im Dienste aller Gebiete, Wissenszweige und Berufe. Zugleich Adreßbuch der Absatzgebiete*. 8th ed. Leipzig: Friedrich Carl Dietze.

Duttenhöfer, Barbara (2013). *Das Geschlecht der Öffentlichkeit. Deutsche und russische Frauenzeitschriften und ihr Publikum im frühen 20. Jahrhundert*. Saarbrücken: universaar.

Gebhardt, Hartwig (1989). Die Pfennig-Magazine und ihre Bilder. Zur Geschichte und Funktion eines illustrierten Massenmediums in der ersten Hälfte des 19. Jahrhunderts. In Rolf Wilhelm Brednich and Andreas Hartmann (eds.). *Populäre Bildmedien. Vorträge des 2. Symposions für ethnologische Bildforschung*, 19–41. Göttingen: Schmerse.

Gesellschaft zur Verbreitung Gemeinnütziger Kenntnisse (ed.) (1833–1842). *Das Pfennig-Magazin für Verbreitung gemeinnütziger Kenntnisse.* Leipzig: Bossange and Brockhaus.

Graf, Andreas (2003). Familien- und Unterhaltungszeitschriften. In Georg Jäger (ed.). *Geschichte des deutschen Buchhandels*, Vol. 1/2, 409–522. München: Saur.

Hanebutt-Benz, Eva-Maria (1984). *Studien zum deutschen Holzstich im 19. Jahrhundert.* Frankfurt am Main: Buchhändler-Vereinigung.

Haug, Christine, and Natalie Kruse (2004). *Geschichte des Versandbuchhandels von seinen Anfängen in den 1860er Jahren bis zur Gegenwart.* Wiesbaden: Harrassowitz.

Hecht, Wilhelm (1897): Allgemeines. In Carl von Lützow (ed.). *Der Holzschnitt der Gegenwart in Europa und Nord-Amerika*, 55–95. Wien: Gesellschaft für Vervielfältigende Kunst.

Illustrierte Weltkriegschronik (1914–1920). Leipzig: J. J. Weber.

Illustrirte Chronik (1848–1854). Leipzig: J. J. Weber.

Illustrirte Zeitung (1843–1944). Leipzig: J. J. Weber.

Illustrirter Kalender (1846–1881). Leipzig: J. J. Weber.

Illustrirtes Weihnachts-Album (1868–1874). Leipzig: J. J. Weber.

Jäger, Jens (2006). Fotografien des globalen Dorfs? Bildjournalismus der 1920er und 1930er Jahre. In Clemens Zimmermann and Manfred Schmeling (eds.). *Die Zeitschrift—Medium der Moderne. Deutschland und Frankreich im Vergleich*, 85–109. Bielefeld: transcript.

Jäger, Jens (2000). *Photographie. Bilder der Neuzeit. Einführung in die Historische Bildforschung.* Tübingen: Edition diskord.

Janzin, Marion, and Joachim Güntner (2007). *Das Buch vom Buch. 5000 Jahre Buchgeschichte.* 3rd. ed. Hannover: Schlüter.

Köhler, Wilhelm (1903). *Preis-Liste über Maschinen, Utensilien, Holzwerk, Materialien etc. für Buchdruck, Stereotypie und Galvanoplastik.* München: Wilhelm Köhler Spezialgeschäft für Buch- und Steindruckereien.

Krüger, Otto (1929). *Die Illustrationsverfahren. Eine vergleichende Behandlung der verschiedenen Reproduktionsarten, ihrer Vorteile, Nachteile und Kosten.* 2nd ed. Leipzig: Brockhaus.

Lachenicht, Susanne (2006). Die neue Visualität der Zeitschrift im frühen 20. Jahrhundert und die culture de masse. In Clemens Zimmermann and Manfred Schmeling (eds.). *Die Zeitschrift—Medium der Moderne. Deutschland und Frankreich im Vergleich*, 63–84. Bielefeld: transcript.

Meisterwerke der Holzschneidekunst (1879–1904). Leipzig: J. J. Weber.

Novellen-Zeitung (1844–1873). Leipzig: Spamer.

Spamer, Otto (ed.) (1858). *Otto Spamer's Illustrationen- & Clichés-Katalog. Verzeichnis der Illustrationen, Vignetten, Initialen [...] von welchen scharfe Clichés in Bleigüssen und Kupferniederschlägen zu den aufgeführten billigen Preisen bis auf Widerruf, stets binnen Kürze zu beziehen sind.* Leipzig: Spamer.

Schmidt, Rudolf (ed.) (1908). J. J. Weber. In Rudolf Schmidt (ed.). *Deutsche Buchhändler. Deutsche Buchdrucker*, Vol. 6, 1025–1028. Berlin and Eberswalde: Verlag der Buchdruckerei Franz Weber.

Schmidt, Rudolf (1908). Spamer, Otto. In Rudolf Schmidt (ed.). *Deutsche Buchhändler. Deutsche Buchdrucker*, Vol. 5, 906–910. Berlin and Eberswalde: Verlag der Buchdruckerei Franz Weber.

Schulze, Friedrich (1926). *Wandlungen des Leipziger Buchgeschmacks in den letzten hundert Jahren. Der Jahresversammlung d. Gesellschaft d. Bibliophilen vom 23. bis 25. Okt. 1926 [...] gewidmet.* Leipzig: Spamer.

Spamer, Otto (1872). *Illustrirter Verlags-Bericht: Jubiläums-Katalog von Otto Spamer in Leipzig*, 2 Abtheilungen. Leipzig: Spamer.

Spamer, Otto (1859). *Vehme oder Justiz? [...] Eine Streitschrift zu seiner Rechtfertigung und im Interesse des gesammten Buchhandels geschrieben.* Leipzig: Spamer.

Stöber, Rudolf (2005). *Deutsche Pressegeschichte.* Konstanz: UVK Verlagsgesellschaft.

Uglow, Jennifer S. (2006). *Nature's Engraver: A Life of Thomas Bewick.* London: Faber and Faber.

Van der Linden, Fons (1990). *DuMonts Handbuch der grafischen Techniken. Manuelle und maschinelle Druckverfahren.* 3rd ed. Köln: DuMont.

Verzeichniss der in Band I–V der Illustrirten Zeitung enthaltenen Abbildungen, von welchen Clichés in scharfen Bleiabgüssen abgegeben werden (1846). Leipzig: J. J. Weber.

Weber, Wolfgang (2003). *Johann Jakob Weber. Der Begründer der illustrierten Presse in Deutschland. Wolfgang Weber.* Leipzig: Lehmstedt.

Weise, Bernd (1989a). Pressefotografie I: Die Anfänge in Deutschland, ausgehend von einer Kritik bisheriger Forschungsansätze. *Fotogeschichte*, 9 (31), 15–40.

Weise, Bernd (1989b): Pressefotografie II: Fortschritte der Fotografie- und Drucktechnik und Veränderungen des Pressemarktes im Deutschen Kaiserreich. *Fotogeschichte*, 9 (33), 27–62.

Weise, Bernd (1990). Pressefotografie III: Das Geschäft mit dem aktuellen Foto: Fotografen, Bildagenturen, Interessenverbände, Arbeitstechnik—die Entwicklung in Deutschland bis zum Ersten Weltkrieg. *Fotogeschichte*, 10 (37), 13–36.

Weise, Bernd (1994). Pressefotografie IV. Die Entwicklung des Fotorechts und der Handel mit der Bildnachricht. *Fotogeschichte*, 14 (2), 27–40.

Weise, Bernd (1996). Pressefotografie V. Probleme zwischen Fotografen und Redaktionen und der Beginn der Bildtelegraphie in Deutschland bis 1914. *Fotogeschichte*, 16 (59), 33–50.

Weise, Bernd (1999). Das Geschäft mit dem Foto in der Geschichte der Presse- und Illustrationsfotografie in Deutschland. In Bundesverband der Pressebild-Agenturen und Bildarchive e.V. (ed.). *Der Bildermarkt. Handbuch der Bildagenturen*, 33–43. Berlin: BVPA.

Weise, Bernd (2003). Aktuelle Nachrichtenbilder "nach Photographien" in der deutschen illustrierten Presse der zweiten Hälfte des 19. Jahrhunderts. In Charles Grivel, André Gunthert, and Bernd Stiegler (eds.). *Die Eroberung der Bilder—Photographie in Buch und Presse (1816–1914)*, 62–101. München: Fink.

Wilke, Jürgen (2013). Journalismus. *Europäische Geschichte Online*, 03/04/2014, http://ieg-ego.eu/de/threads/europaeische-medien/journalismus/.

Wustmann, Gustav (1893). Spamer. In Historische Kommission bei der Bayerischen Akademie der Wissenschaften (ed.). *Allgemeine Deutsche Biographie (ADB)*. Vol. 35, 31–32. Frankfurt am Main: Duncker & Humblot.

Zierenberg, Malte (2013). Die Ordnung der Agenturen. Zur Verfestigung massenmedialer Sichtbarkeit im Pressewesen, 1900–1940. In Annelie Ramsbrock, Annette Vowinkel, and Malte Zierenberg (eds.). *Fotografien im 20. Jahrhundert. Verbreitung und Vermittlung*, 44–65. Göttingen: Wallstein.

Eyewitnesses? The Visual Depiction of Events around 1900

Jens Jäger

The sinking of the *Titanic* made big news in April 1912. Wireless and telegraphic communication transmitted the fact that the steamer had sunk and had sent out distress signals. Newspapers and magazines immediately reported the event, gave background information, provided human-interest stories, and mused on the consequences of such a disaster. Numerous periodicals published images of the ship, the passengers, and the crew. Maps showed the course of the ill-fated steamer, and charts and diagrams added details. There was extensive coverage of the incident in words and images even though no first-hand reports were available until after survivors reached the next harbor (April 19), and there were no images of the incident itself: no photographs, no sketches made on the spot. Nevertheless, newspapers and magazines suggested that their reporting also unfolded the whole story visually. This raises questions about the visual depiction of the shipwreck: Where did the published images come from? How were they used and for what purposes?

From today's perspective it seems self-evident that media coverage of an event such as the shipwreck of a large passenger vessel includes text and images. However, around 1900 it was uncommon for many daily newspapers to publish images. Even magazines had difficulties obtaining what we would today call up-to-date topical images, especially photographs, which in 1912 were not (yet) used to report a topical occurrence. The reasons for this vary: Photography was not yet sophisticated enough to cater to this need, and photographers were by far not as ubiquitous as they would become in later decades. Furthermore, printing images in newspapers or magazines was expensive, and many editors of daily newspapers abstained from providing visual information because it did not fit the concept of the paper, for instance *The Times.* On the other hand, the illustrated press was already well established by the turn of the twentieth century, and the reading public was accustomed to seeing images in magazines and books.

Surprisingly, until recently research on the history of visual depictions in the periodical press rarely touched on theses questions. Although photojournalism has received increased attention in the past twenty years, comprehensive critical histories of press photography have yet to be written (Weise 2012, 63; Barnhurst 1994). The same applies to the study of press illustration. Browsing through illustrated periodicals of the nineteenth and early twentieth centuries in Europe and the Americas, two phenomena become apparent. First, many types of images were in use: the woodcut, wood engraving, various forms of metal engraving, lithography, and—from the eighteen-eighties onwards—photography. The latter did not replace other means of illustration; rather, it was "used simply as one tool among many" (Barnhurst and Nerone 2000, 60). Second, illustration practices and technologies were applied worldwide, and the images published therefore look surprisingly similar (although not necessarily the motives). The market for news illustrations was never restricted to limited areas, in particular around 1900 when news and picture agencies began gaining more influence in the market (Zierenberg 2013). Again, a lack of research on this topic limits the scope of the following considerations.

The ensuing paragraphs discuss the way images were employed in periodicals around 1900. The news coverage of the sinking of the *Titanic* in April 1912 serves as an example. This may not be the best case in point for demonstrating how the press visually depicted events around 1900 in everyday practice, as three things guaranteed journalistic and public awareness: It was the biggest and most modern passenger vessel of the time, it claimed the hitherto highest number of lives, and it involved a considerable number of Americans and Europeans. There was therefore no need for spectacular images to gain attention. However, it is a good example for illustrating that combinations of different types of images were very important in the periodical press at the time. As far as the use of pictures of any kind in the press is concerned, the turn of the twentieth century was a time of transition: Photography had become increasingly important in newspapers and was already a major factor in magazines. Engravings still played an important role in the reporting of events—more than usually acknowledged in recent research (Barnhurst and Nerone 2000, 66). Charts and diagrams provided additional information. All in all, it was images, the prominent space in the papers dedicated to the story, and the ongoing coverage over several weeks that marked an occurrence as an important (global) event.

While it is easy to assert that images provide a different kind of information than words, it is more difficult to explain the function of images within the periodical press during a given period in time, as they are complex and cannot be reduced to mere visual evidence. For example, photographs published in conjunction with an event did not necessarily represent what eyewitnesses saw. However, they were important for drawing reader attention, be it at the bookstall or while the readers turned the pages. Although photography as a technology was not new, it still was innovative as a means of illustration in many daily newspapers.

Before turning to the concrete example of the *Titanic,* it is necessary to briefly discuss several questions that enable better understanding the function and use of images in the periodical press around 1900. What was the intention of editors in providing visual information? Were did the images come from? How important was the technique used—photography, diagrams, maps, charts? Was photography used as a medium to highlight a reported occurrence as an event and to construct the idea of "eyewitnessing"? An eyewitness is someone who has personally seen something take place, which implies that he or she was physically present for a certain period of time. Newspaper readers looking at a printed photograph are not consistent with such a definition, but editors suggested otherwise, as will be shown later. How important were images for the fabrication of "media events," which, I contend here, are to a considerable extent driven by the way they are depicted in the (daily) press? The news coverage of the *Titanic* disaster in April 1912 is a case in point. Although it would seem that this incident was an "event" that did not require the fabrication of an interesting story, the press coverage provides insight into the mechanisms of making news and transmitting it by means of images.

At this point it is useful to briefly discuss the term "event." Frank Bösch recently suggested being very careful with the term and differentiating between certain types of "events." However, whether we speak about "historical," "pseudo" or "media-driven" events, what they all have in common is that media are crucial to their constitution (Bösch 2010). An occurrence like the sinking of a ship or a politically motivated murder is not an event per se. Bösch asserts: "In the modern period, the broad communication necessary to transform occurrences into events takes a medial form and occurs in a supra-regional space. The public construction of meaning—whether performed by messengers, pamphlets, or television—and its attributions act as the mechanism that transforms an occurrence into an event." According to

Bösch, the driving force behind creating media events in the late nineteenth century is undoubtedly to be found in the context of the emerging or growing mass media market. Media needed content, and they needed content fit for reporting. A well-known example is the *Tour de France* initiated by the magazine *L'Auto* in 1903. It is no coincidence that such events were invented during a period that was characterized by a proliferation of images in the media. Images created awareness or underlined the importance of the occurrence being covered, and they distinguished it from other stories in the news. This works the other way around: What was deemed important also had to be accompanied by its visual depiction. The decades around 1900 were pivotal in this respect.

Images in the Press

With the introduction of pictures and illustrations in magazines in the eighteen-thirties, images served as information as such, supplementary information, visual evidence of something, corroboration of the facts, and as an invitation to readers to scrutinize the accuracy of the news being reported for themselves. In the words of the editor of the *Illustrated London News* in 1842: "The public will have henceforth under their glance and within their grasp the very form and presence of events as they transpire" (*Illustrated London News* 1842). This short sentence includes two important assertions: First, that readers become eyewitnesses, and second, that what they see in the press is an "event" or part of one. This suggested to readers that they were present when something took place, which subliminally added an emotional dimension highlighted by the words "glance" and "grasp." A decade later, Ernst Stolle and Ernst Keil (1853), who introduced the German illustrated magazine *Gartenlaube,* contended that images would not only inform readers but also augment the magazine by an aesthetic dimension. They promised that the *Gartenlaube* would "publish appealing and explanatory pictures by renowned artists. Thus, we want to entertain and while doing so instruct you in an entertaining fashion" (Stolle and Keil 1853, 1). By the mid-nineteenth century, at least six major functions of images in the periodical press can be identified: They serve as substitutes for being present, they instruct, they represent something that ought to be perceived as an "event" or at least as "important," they adorn the content, and they emotionalize and entertain

(Barnhurst and Nerone 2000, 64). Another intended effect is, of course, to attract readers.

Given the importance of images, it is not surprising that a market for visual information developed from the mid-nineteenth century onwards. In the early years, editors commissioned or employed artists and draughtsman or relied on established picture-sellers. The exchange of material from publisher to publisher constituted another source (Gebhardt 1989; Lachenicht 2006, 66). Some publishing houses encouraged the public to submit their own pictures; however, this turned out to be insufficient. Over the decades, the majority of publishers, such as—to give just a German example—Brockhaus, Scherl, Ullstein, and Mosse, assembled their own stocks of images. The *New York Times* also had an extensive and well-organized archive. All of the others had to fall back on external sources. In the late nineteenth century, specialized agencies became more and more important for meeting the growing press sector's demand for images (Zierenberg 2013, Knoch 2006). Specialized stock photo agencies included the Berliner Illustrations-Gesellschaft (est. 1900) and Robert Sennecke (est. 1910) in Berlin, Henri Manuel (est. 1900) and Meurisse (est. 1905) in Paris, and the Topical Press Agency (est. 1903) and the Sport and General Press Agency (est. before 1910) in London. The American agency Keystone was founded as early as in 1891. It was common for newspapers and magazines to use stock images or pictures provided by agencies to illustrate a story. This at least partially accounts for the similarity of visual information in the press, as can be shown in the example of the *Titanic*.

Generally, a wide range of subjects was depicted visually: people, architecture, art (reproductions), scientific discoveries, landscapes, nature, and incidents of all sorts, including catastrophes. What seemed to be a comprehensive representation of the world was in fact a highly selected view of the natural and cultural environment that catered to the (supposed) demand of discrete target groups of the press. Images published in newspapers and magazines had been subjected to a long selection process. It took time to obtain or produce pictures, which meant that they were rarely direct visual documentation of current events but reproductions based on how artists and editors perceived them to be.

Most images published provided more background and supplementary information (often portraits of people mentioned in the text) than first-hand reports. They often depicted the outcome of an incident or suggested a decisive moment according to the principles of historical and genre painting.

The expression "decisive moment" later became proverbial for the work of news photography following the publication of Henri Cartier-Bresson's *The Decisive Moment* in 1952. This should not be confused with earlier concepts of press photography. Until the nineteen-twenties, the bread-and-butter work of a photojournalist consisted of nontopical, technically perfect, "interesting," or "original" images (Boedecker 1926). Drama and action were seen as exceptions and not something on which a press photographer should base his career. The contradiction to the claim made by the *Illustrated London News* mentioned above is obvious. Images in the press could not "grasp the very form and presence of events as they transpire"; rather, they staged, interpreted, analyzed, or dramatized them. In other words, images were essentially a means of reflection. Their function was more in line with that suggested by Stolle and Keil: to blend entertainment with instruction. Photographs did not really alter this attitude toward images.

The alleged basic functions of images in the periodical press did not change with the introduction of the halftone process in the early eighteen-eighties. Photographs reinforced the idea that objects or occurrences reported by this technical means provided more objective information than any other form of visual representation. They had therefore already served as templates for engravings since the eighteen-forties (Weise 2003). If a camera was used there was no reason to cast any doubt on the authenticity of an image. Photographs seemed to register more than the photographer had been aware of; minute details invisible to the "naked eye" could be discerned in a picture. Photography was furthermore used as evidence and documentation in science, was applied for administrative purposes, and a growing number of people had first-hand experience with the technique, either as amateurs or as clients of professional photographers. Regardless of whether photographs were staged or how often they were retouched, if the source was reliable then the image itself was also considered to be reliable.

Although photography did not fundamentally alter the way images were used in the press, they reinforced the aspect of eyewitnessing and objective reporting (Weise 2003, 74–75). They were applied less to "adorn" and more to make newspapers and magazines more appealing, the news seemingly more reliable. Editors were alert to the fact that using the most advanced technologies available promoted the modernity of their paper or magazine.

It is important to note, however, that although hand-held cameras were in use, many photojournalists still employed large-format cameras on tripods up to the nineteen-twenties (Boedecker 1926, 13). Glass plates remained much more important than roll film. In short: For technological reasons, most news photography was more static than dynamic and less up-to-date in the sense of capturing unanticipated "action." Static images were easier to control as far as motive and aesthetics were concerned. The primary task of photography was not to simply report dramatic events but to document the consequences, the background, or the outcome. This was different for carefully orchestrated events, such as recurring celebrations, public holidays, anniversaries, sports, and all kinds of shows and exhibitions. Another aspect needs to be mentioned: Although photography was a "fast" medium in terms of recording, it took time for the photographer to develop the negative and produce the prints to be forwarded to an agency or editorial office. It was a week or more before images were published. Of course, daily newspapers could react more rapidly than weekly publications, but around 1900 it was unlikely—although possible—for a commissioned photograph to appear within twenty-four hours. According to Christian Delporte, who researched the French magazine *Excelsior*, it normally took a week before an image could be published (Lachenicht 2006, 67). Many magazines used stock images to illustrate unexpected topical events. This is one reason why postcards depicting such events (catastrophes, accidents) were in a position to compete with newspapers: Photographers could produce a large number of them overnight, far in advance of other media.

After a brief period of reluctance, more and more illustrated magazines (and newspapers) began making increased use of photographic illustrations. This was not a technological or economical necessity but indicated a gradual change in the "visual culture" of the press (Barnhurst and Nerone 2000, 59), which, however, will not be discussed here. Yet it is important to note that numerous periodicals still published non-photographic illustrations even after the introduction of photographs, including the European and American publications *Berliner Illustrierte Zeitung* (est. 1892), *Die Woche* (est. 1899), the *Illustrated London News* (est. 1842), *L'Illustration* (est. 1843), *Le Monde illustré* (est. 1857), and *Harpers Weekly* (est. 1857), to name only a few. Daily newspapers also began to include photographs, despite the fact that the halftone process was considered too complex and expensive a process for a daily. The English *Daily Graphic* (est. 1890) therefore concentrated on engravings. The *B.Z. am Mittag* (est. 1904), one of the large German daily newspapers, employed draftsmen; prior to 1918, it rarely, if at all, included photographs (Berliner Zeitung 2012). Even dailies with a very large circulation, such as *Le Petit Parisien* (est. 1876, by 1900 allegedly one of the newspapers with the largest circulation in the world), *L'Intransigeant* (est. 1880), or *Le Matin* (est. 1884), did not begin to use photographs until after 1900. The evidence does not support the notion of a "halftone revolution." However, others saw an opportunity in the new medium and invested considerable capital: The English *Daily Mirror* (est. 1903/04) and the French newspaper *Excelsior* (est. 1910), for example, were lavishly illustrated with photographs.

Prior to World War I, images never dominated the spaces of daily newspapers (Weise 2003, 101), although there were exceptions, such as the *Excelsior* or the *Daily Mirror* (Guéry 2006, 144). Their increased use put pressure on the competition. To keep pace with the *Excelsior,* other French newspapers felt obliged to include or increase their volume of photographic images. However, newspapers such as the German *B.Z. am Mittag,* the *Vossische Zeitung,* the *Kreuzzeitung,* or Austria's third-largest newspaper, the liberal *Neue Freie Presse,* continued their policy of no images, except in the advertising section, without suffering a loss of sales.

It is safe to conclude that images were increasingly used in the periodical press around 1900, although only sporadically to illustrate (highly) topical events. To attract readers, they stressed the authenticity of the news, added interesting subjects and emotional qualities, and adorned the pages. They provided more detailed visual information in the mode of "eyewitnessing"—as stressed by contemporary editors and photojournalists—without actually conforming to the definition of "eyewitness." Like other types of images in daily newspapers, photographs demonstrated the importance and reliability of the news. In most cases, portraits were used that were readily available in the archives or collections of agencies (Zierenberg 2013). Photographs were often published in magazines. Yet they could not completely replace engravings, because "action" was very difficult to record photographically, except in highly controlled environments such as those described above. Though photographic images enhanced the importance or relevance of a story, entertainment played a role as well.

The Sinking of the Titanic—Visual Depictions of an "Invisible" Event

By 1912, photographs were well established in the press, common in magazines, and becoming increasingly important in daily newspapers. When the news arrived by wireless that the largest steamer in the world had collided with an iceberg on its maiden voyage, editors immediately realized that written accounts of this event were not sufficient in and of themselves. An overview of 36 front-pages of American, British, French, and Canadian newspapers published between April 15 and 28, 1912, indicates that most editors were eager to add visual information to their articles (Bossi 2013). Photographs of the vessel that had already been in circulation since it was launched in 1911, went on sea trials in April 1912, and left for its maiden voyage were now used to illustrate the articles.

Magazines had more time to create their own "image" of the disaster. By 1912, the editors of the *Gartenlaube* knew only too well that breaking news required images. The space dedicated to images in the magazine had therefore grown from about 17 percent in 1875 to 43 percent in 1885, and by 1895 it had reached 45 percent (Wildmeister 1998). The sinking of the *Titanic* was an occasion to employ more than just the stock photographs taken when the steamer had left Belfast or Southampton earlier in April. The editors quickly commissioned the German artist Willy Stöwer (1864–1931) to render a version of the catastrophe after the news arrived that the steamer was lost. Stöwer was already well known at the time for his nautical paintings and lithographs. His drawing of the event was published on Thursday, May 2, 1912 in issue no. 19 of the *Gartenlaube* (Fig. 1). The artist imagined the perspective of a survivor in one of the lifeboats. He deliberately changed aspects of the incident, partly because he may not have known better (the

Fig. 1: Willy Stöwer (1864–1931), Der Untergang der Titanic (wood-engraving) (Source: Die Gartenlaube, May 1912)

rear funnel of the *Titanic* was for decoration purposes only and would not have billowed smoke) and partly to dramatize the scene. His image was to become iconic and has since inspired numerous other visual representations of the *Titanic's* demise (Helmes 2009).

It was obvious that no one would have been able to take a picture of the sinking of the ship—the *Titanic* collided with the iceberg close to midnight and went down at approximately 2:30 in the morning. The photographic equipment of the time would have been incapable of capturing the incident in the middle of the night. Even if someone had successfully taken a photograph, the steamer *Carpathia* with the survivors on board did not reach New York until April 18, the earliest possible date for the publication of an original photograph thus being April 19. Nevertheless, most editors of newspapers and magazines immediately sought to obtain photographs of the *Titanic,* its passengers, its crew, or icebergs; they ordered maps, diagrams, and charts. Some commissioned illustrators or painters to depict the disaster.

In 1912, on such an occasion the editors of newspapers and magazines opted for images that they and the public believed to be objective and trustworthy: photographs or engravings based on photographs, or at least engravings based on drawings from a reliable source. The *Gartenlaube* decided for the latter, commissioning a celebrated painter of nautical subjects to depict the event. Certain elements that were known by the time the image was published had to be included: the iceberg (which in fact was not even near the position where the *Titanic* foundered), passengers in the freezing water, the moment the stern lifted clear out of the water before the ship finally went down. That was of course a very dramatic moment with a strong emotional quality.

How did other editors react? Daily newspapers made the news the leading article. Those who had the opportunity added visual information, in most cases a photograph of the steamer or portraits of the captain and prominent passengers. Yet whatever the case, the public did not have "under their glance and within their grasp the very form and presence of events as they transpire," as the editors of the *Illustrated London News* had boasted seventy years earlier and which was still claimed by contemporary editors.

Those newspapers capable of publishing photographs did so immediately after the news of the steamer's demise arrived: On April 16, 1912, *Le Petit Parisien* featured a photograph of the *Titanic* on the front page (Fig. 2), as did the *New York Times* (New York), *Le Matin* (Paris), and numerous others in Europe and the United States; the *Daily Mirror* (London) included a photo-

Mardi 16 Avril 1912

Le Petit Parisien

5 centimes — Le plus fort tirage des journaux du monde entier — 5 centimes

ÉDITION DE PARIS

Pauvres Poètes !

POLITIQUE NATIONALE

On retrouve le ballon « Zodiac » en plein Océan

M. Poincaré parle

L'ÉMOUVANT ACCIDENT DU « TITANIC »

Le plus grand paquebot du monde heurte un iceberg devant Terre-Neuve

Mais on a pu, heureusement, sauver les 2.358 personnes qui se trouvaient à bord

« LE TITANIC »

LE DANGER DES ICEBERGS

A CHERBOURG

UN DRAME A CHATILLON

Ayant assommé sa femme, un mari danse sur elle

Scène de carnage

Un forcené

L'ECLIPSE

DEMAIN, LA LUNE CACHERA LE SOLEIL

Ce que l'on verra

Fig. 2: Le Petit Parisien (April 16, 1912), Front Page

The Daily Mirror

THE MORNING JOURNAL WITH THE SECOND LARGEST NET SALE.

No. 2,645. TUESDAY, APRIL 16, 1912 One Halfpenny.

DISASTER TO THE TITANIC: WORLD'S LARGEST SHIP COLLIDES WITH AN ICEBERG IN THE ATLANTIC DURING HER MAIDEN VOYAGE.

Disaster, it was reported yesterday, has overtaken the great steamer Titanic, the largest and most luxuriously appointed vessel afloat. The liner, which is the latest addition to the White Star fleet, left Southampton last Wednesday on her maiden voyage to New York, and was in the vicinity of the Newfoundland banks, to the south of Cape Race, when she struck an iceberg, an ever-present peril in those latitudes at this time of the year. "Wireless" has again demonstrated its immense value, assistance being summoned by this means. The photograph shows the mighty vessel leaving Southampton on Wednesday.—(*Daily Mirror* photograph.)

Fig. 3: The Daily Mirror (April 16, 1912), Front Page

graph of the steamer that covered four-fifths of the page (Fig. 3). In contrast, the *B.Z. am Mittag* (Berlin) and the *Neue Freie Presse* (Vienna) stuck to their policy and ran "only" articles. *The Times* (London) published a map with the position of the *Titanic* and other steamers, such as the *Olympic*, the *Virginian*, and the *Baltic*, at the time the distress signal was sent. Since the first pages of the venerable London newspaper were always dedicated to advertisements, the map appeared on page ten alongside an article with details about the incident and included a list of the first-class passengers (Fig. 4).

The April 16 issue of the *New York Herald* featured an engraving depicting the ship at night steaming (full speed?) through an agitated sea toward an iceberg (Fig. 5) and around which photographs of passengers and the crew and a small map were arranged in a semicircle. The engraving must have been produced remarkably fast; it is also possible that a dummy had been prepared in advance. The paper opted for drama, obviously sensing an opportunity to upstage the other New York papers. The *New York Herald's* emphasis on immediacy and emotional impact was thus in line with a trend in North American illustrated journalism as detected by Barnhurst and Nerone (2000). This presumably accounts for the difference compared with European newspaper reports on the catastrophe.

Apart from this distinctive feature of North American journalism, the *New York Herald's* coverage of the story indicates that drama was still best expressed and emotions best aroused by means of engravings. Charts and maps provided additional information, and portraits added a human face. The action represented was in fact an artist's interpretation based on the requirements laid down by journalists and editors in response to supposed reader/viewer demand—the latter were thus encouraged to imagine themselves as having actually eyewitnessed the disaster, as the engraving in the *Gartenlaube* suggested. Other periodicals followed suit: On April 21, 1912, just under a week after the incident, Vienna's tabloid *Wiener Bilder* featured a drastic image on its front page showing a huge wave surging over the deck of the *Titanic* (Fig. 6). The April 28 issue of the *Berliner Illustrierte Zeitung* contained an allegorical impression of the incident showing the *Titanic* steaming directly toward a gigantic iceberg with a skull at its peak. On the same day, the Sunday supplement of the *Petit Journal* featured a colorful engraving of the *Titanic* colliding with the iceberg (Petit Journal 1912, 136).

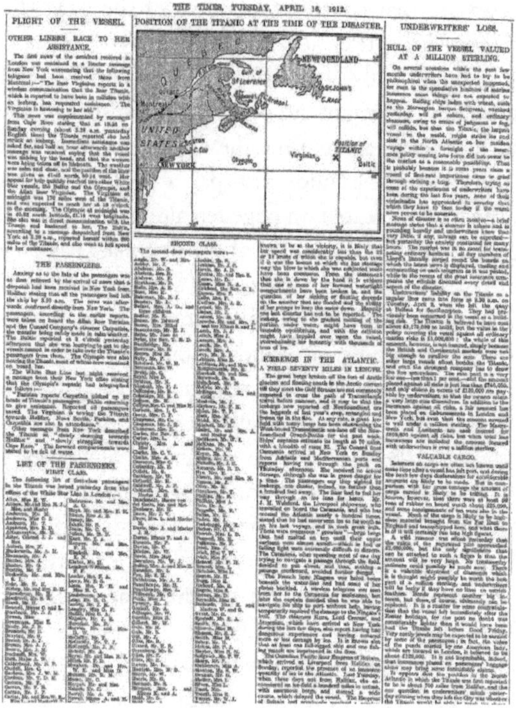

THE TIMES, TUESDAY, APRIL 16, 1912.

FLIGHT OF THE VESSEL.

OTHER LINERS RACE TO HER ASSISTANCE.

THE PASSENGERS.

LIST OF THE PASSENGERS.
FIRST CLASS.

POSITION OF THE TITANIC AT THE TIME OF THE DISASTER.

SECOND CLASS.

ICEBERGS IN THE ATLANTIC.
A FIELD SEVENTY MILES IN LENGTH.

UNDERWRITERS' LOSS.

HULL OF THE VESSEL VALUED AT A MILLION STERLING.

VALUABLE CARGO.

Fig. 4: The Times (April 16, 1912) Page 10 (excerpt)

The sinking of the *Titanic* became more than just a maritime catastrophe. For a long time, it was *the* paradigmatic shipwreck. Worldwide press coverage in words and pictures over several weeks caused the occurrence to become a global event (and a media event as well). As such it entered universal cultural memory. Contemporary images of the catastrophe itself were quickly rendered fantasies by draftsmen and artists yet reflected a complex and well-established mode of illustrated journalism and informed the collective visual memory of the event for decades—much more so than articles or commentaries.

THE NEW YORK HERALD.

THE TITANIC SINKS WITH 1,800 ON BOARD; ONLY 675, MOSTLY WOMEN AND CHILDREN, SAVED

MOST APPALLING DISASTER IN MARINE HISTORY OCCURS WHEN WORLD'S LARGEST STEAMSHIP STRIKES GIGANTIC ICEBERG AT NIGHT

Scores of World's Most Widely Known Persons, Including Colonel John Jacob Astor and His Wife, William T. Stead and Isidor Straus Among Those Whose Fate Is in Doubt.

J. Bruce Ismay, Head of the White Star Line; Mr. and Mrs. George D. Widener, the Countess of Rothes and Mrs. Henry B. Harris Are Some of Those Saved.

Fig. 5: The New York Herald (Aptil 16, 1912), Front Page

Eine Rundfrage an die bekanntesten Wiener Verteidiger: „Mein interessantester Prozeß“

V. Chiavacci's

Wiener Bilder

Illustrirtes Familienblatt

Redaktion und Administration Wien III. Rüdengasse 11. Telephone: Nr. 9767 Nr. 4199 Nr. 9540

Nr. 16. Wien, Sonntag, 21. April 1912. XVII. Jahrgang.

Abonnements-Bedingnisse: Für Oesterreich-Ungarn vierteljährig K 2·50, halbjährig K 5·—, ganzjährig K 10·—; für Deutschland vierteljährig M. 3·50, ganzjährig M. 14·—; für alle übrigen Weltpostvereinsländer [illegible] — Inseratenannahme durch alle Annoncenbureaus des In- und Auslandes. — [illegible]
Redaktion und Administration: III. Rüdengasse Nr. 11. — Telephone: Nr. 9767, 4199, 9540. — K. k. Postsparkassen-Konto Nr. 54.154. — Einzelne Nummern 20 Heller.

Die größte Schiffskatastrophe aller Zeiten.

Zum Untergang des „Titanic“: 10 Bilder.

Die neuen städtischen Gaswerke. ♦ Neue Pariser Hutmodelle. ♦ Englische Fußballspiele in Wien.

Fig. 6: Wiener Bilder, 17 (16) (April 21, 1912), front page

The importance of the incident was highlighted by the combination of text and images. Photographs underscored the authenticity of what was being reported, and the reports corroborated the images. Charts and maps provided additional facts and background information, thus confirming the reliability of the words and pictures. Portraits of passengers, the crew, and next of kin made the incident "human" and credible. The different types of images vouched for the reliability of each image and the written accounts. In the case of the *Titanic,* stock images were combined with images produced in the wake of the disaster. The emotional and dramatic core of the story (the sinking of the ship) was visually depicted by means of engravings.

Up to the nineteen-twenties, photojournalists rarely covered incidents, catastrophes, accidents, and the like. This was due to the technical restrictions of photography, which also affected printers and publishers. Speed was important, and so to compensate for the relatively lengthy process required to publish images in the periodical press, a well-oiled trade in images supplied visual source materials ready for printing. While this material was generally not up-to-date, it was nevertheless presented as something close to eyewitnessing.

Around 1900 it became more and more important to supplement the coverage of topical events or at least major occurrences with images, especially photographs. The presence of an image in itself indicated the importance of a story. Photographs accounted for the fidelity of reporting—the more reliable the image by contemporary standards the better. This "reliable" material backed up contrived images, such as engravings that depicted an occurrence. Engravings that provided a rendering of the fate of the *Titanic,* for example, were more important to collective memory than written accounts. Images circulated longer and were used time and again, as they remained available through archives and agencies. Stöwer's engraving became "iconic"; photographs of the ship, the crew, and survivors continue to circulate to this day and can be acquired from stock photo agencies worldwide (for instance Corbis, Getty Images, AP Images, Süddeutsche Zeitung Photodienst, to name but a few).

Conclusion

Research in the areas of press illustration and photojournalism no longer centers on any sort of technological determinism. Economic conditions and the organization of the trade in images are also important. The reports on the *Titanic* disaster in April 1912 demonstrate that most newspapers were in a position to illustrate the story. Although at the turn of the twentieth century words traveled much faster (telegraphy, wireless, telephone) than images, editors and journalists could narrow the speed gap. They included illustrations that in themselves were not directly connected to the incident. Many types of images were employed and suggested the exhaustive reporting of a global event. Stock images from the archives of publishers and image agencies guaranteed that occurrences of any importance were accompanied by an illustration. Illustrators and engravers worked efficiently if exclusive material was needed. However, some successful newspapers could do without visual information. If and what kind of images were published depended far more on traditions of (visual) journalism and cultural developments. These traditions seem to be—at least in part—transnational, even though the cultural context may have differed. Photographs highlighted objectivity and fidelity. Engravings were a means to report, emphasize, and emotionalize, and they added interpretation and narration. They represented what could not be photographed yet claimed to be just as authentic or reliable. Maps and diagrams completed the set of illustrations, supplying matter-of-fact details. This combination of images was effective at least up to World War I. Readers (and viewers) could imagine themselves as being "eyewitnesses" so far as they were already accustomed to this form of visual information. They witnessed what journalists sold as a comprehensive narrative of an occurrence in word and images. Newspapers and magazines included photographs that could be read as proof of what was written or shown in other images. That by no means implies that the reading and viewing public around 1900 was naïve or uncritical; however, the lasting success of this mode of informing indicates that it measured up to expectations—and the basic principles are still valid today: Different types of images (still or moving) in combination with text (written or spoken) configure the narrative of reports. The amount of image- and text-based information indicates the assumed importance of an occurrence, especially if a topic "makes news" over a considerable period of time. The *Titanic* disaster is a case in point. However, this should not obscure the fact that the critical history of visual journalism or the illustration of newspapers and magazines has yet to be written, and that numerous aspects of this history have to be examined from a global or transnational perspective.

Works Cited

Barnhurst, Kevin G. (1994). Photography as Culture: Reconsidering the History of Photojournalism. *Medien & Zeit,* 1, 17–24.

Barnhurst, Kevin G., and John Nerone (2000). Civic Picturing vs. Realist Photojournalism: The Regime of Illustrated News, 1856–1901. *Design Issues,* 16, 59–79.

Berliner Zeitung (2012). Das wichtigste Thema blieb gleich: Berlin. *BZ online,* October 19, 03/20/2013, http://www.bz-berlin.de/135-jahre/das-wichtigste-thema-blieb-gleich-berlin-article1565940.html.

Boedecker, Ludwig (1926). *Pressephotographie und Bildberichterstattung. Ein Handbuch für Pressephotographen.* Bunzlau: Verlag "Der Photograph" L. Fernbach.

Illustrated London News (1842). *Leading Article.* Vol. 1, no. 1, 1.

Bösch, Frank (2010). European Media Events/Europäische Medienereignisse, *European History Online/Europäische Geschichte Online 2010,* 02/21/2014, http://www.ieg-ego.eu/de/threads/europaeische-medien/europaeische-medienereignisse.

Bossi, Claudio (2013). *Cosa scrissono i giornali,* 03/19/2013, http://www.titanicdiclaudiobossi.com/Html/Cosa%20scrissero%20i%20giornali_151.htm.

Cartier-Bresson, Henri (1952). *The Decisive Moment.* New York: Simon & Schuster.

Gebhardt, Hartwig (1989). Die Pfennig-Magazine und ihre Bilder: Zur Geschichte und Funktion eines illustrierten Massenmediums in der ersten Hälfte des 19. Jahrhunderts. In Rolf Wilhelm Brednich and Andreas Hartmann (eds.). *Populäre Bildmedien: Vorträge des 2. Symposions für ethnologische Bildforschung. Rheinhausen bei Göttingen 1986,* 19–41. Hamburg: Schmierse Media.

Guéry, Louis (2006). *Visages de la presse: Histoire de la présentation de la presse française du XVIIe au XXe siècle.* Paris: Editions Victoires.

Helmes, Günter (2009). Der Untergang der Titanic: Modellkatastrophe und Medienmythos. In Gerhard Paul (ed.). *Das Jahrhundert der Bilder. Vol. 1: 1900–1949,* 124–131. Göttingen: Vandenhoeck & Ruprecht.

Knoch, Habbo (2006). Living pictures: Photojournalism in Germany 1900 to the 1930s. In Karl Christian Führer and Corey Ross (eds.). *Mass Media, Culture and Society in Twentieth-Century Germany,* 217–233. Basingstoke: Palgrave Macmillan.

Lachenicht, Susanne (2006). Die neue Visualität der Zeitschrift im frühen 20. Jahrhundert und die culture en masse. In Clemens Zimmermann and Manfred Schmeling (eds.). *Die Zeitschrift—Medium der Moderne: Deutschland und Frankreich im Vergleich,* 63–84. Bielefeld: transcript.

Petit Journal: Supplement illustré (1912). Vol. 23, 129–136.

Weise, Bernd (2003). Aktuelle Nachrichtenbilder "nach Photographien" in der deutschen illustrierten Presse der zweiten Hälfte des 19. Jahrhunderts. In Charles Grivel, André Gunthert, and Bernd Stiegler (eds.). *Die Eroberung der Bilder: Photographie in Buch und Presse 1816–1914*, 62–101. München: Wilhelm Fink.

Weise, Bernd (2012). Zeit im Bild. Über Quellen und Recherchen zur Mediengeschichte der Pressefotografie. *Fotogeschichte*, 124, 62–66.

Wildmeister, Birgit (1998). *Die Bilderwelt der Gartenlaube: Ein Beitrag zur Kulturgeschichte des bürgerlichen Lebens in der zweiten Hälfte des 19. Jahrhunderts*. Würzburg: Bayerisches Nationalmuseum.

Zierenberg, Malte (2013). Die Ordnung der Agenturen: Zur Verfertigung massenmedialer Sichtbarkeit im Pressewesen, 1900–1940. In Annelie Ramsbrock, Annette Vowinckel, and Malte Zierenberg (eds.). *Fotografien im 20. Jahrhundert*, 44–65. Göttingen: Wallstein.

Mobile, Networked and Digital Technology: Implications for Journalistic Work

John V. Pavlik

Digital Technology Trends

Since the advent of mass media in the 19th century, journalistic work has relied increasingly on the use of technology. With the invention of the Daguerreotype, for instance, Samuel Morse recognized the value of photographic imagery in newspaper reporting. Morse advised his brother, Sidney, then editor of the *New York Observer*, on the great potential in this new technology. Having witnessed for himself this new technology while in Paris, Morse described the Daguerreotype as "one of the most beautiful discoveries of the age." Sidney published his brother's description in the *New York Observer* on April 20, 1839, and it "was widely copied by other newspapers through-out the country" (Morse 1839). Morse taught Matthew Brady and others the technique of the Daguerreotype and soon newspapers were publishing photos of a wide variety of subjects. Poet Walt Whitman, then editor of the *Brooklyn (NY) Eagle* newspaper, himself observed the value of photographic imagery in storytelling. Of the Daguerreotype portrait he wrote in the July 2, 1846 edition of his paper, "*In whatever direction you turn your peering gaze, you see naught but human faces! There they stretch, from floor to ceiling—hundreds of them. Ah! what tales might those pictures tell if their mute lips had the power of speech! How romance then, would be infinitely outdone by fact*" (Whitman 1846).

In the past century, journalists have used a variety of other technologies for newsgathering and storytelling. In the 20th century journalists working in electronic media such as radio and television relied on microphones and cameras to gather news. With the development of digital technology, journalists of all types have used computers for a variety of purposes, such as to write and edit their stories, whether text-based of multimedia, or to analyze vast collections of data, mining them for stories. The increasing miniaturization of digital devices driven by Moore's Law has lead journalists of all types in the 21st century to rely on a variety of mobile and other devices.

The Internet and other digital technologies have given reporters access to an increasing array of online tools and resources. These include Google Earth to provide mathematically accurate maps and online public records that enable producing stories based on so-called Big Data and data-driven visualizations.

Ithiel de Sola Pool (1983) once observed that for five hundred years battles were fought around the world to guarantee the right to speak and print freely and openly. As electronic media grew in prominence, Pool explained, they did not inherit the same "unabridged right of citizens to speak without controls." They were not "technologies of freedom." The situation has seemingly worsened as digital technologies have taken center stage in the 21st century and governments around the world have acted to monitor, censor and otherwise control citizens' use of digital media.

There are serious problems facing bloggers, tweeters and other citizen journalists or networked individuals participating in online discourse around the world. From *'lèse-majesté' (injured majesty) laws* to crackdowns on freedom of speech online, hundreds of bloggers and users of Twitter have been jailed in the past year for seemingly minor speech offenses in many countries. From Kuwait to Oman to Vietnam, the social networking space is increasingly under government control. A Kuwaiti court in March 2013 sentenced Rashid al Enizi to two years in jail for tweeting information deemed insulting to the country's Emir. His tweet: "The coward has run away after issuing an order." The threat of censorship and other forms of government repression are creating a chilling effect on digital journalism, especially citizen journalists, who lack institutional support.

Yet, a shift in the basic paradigm of how journalism works is underway, driven by recent technological developments. This shift presents the possibility of greater openness and less control in the digital creation and flow of news, if not in the governmental regulation of such digital discourse.

A New Paradigm of News Reporting

In the past decade, a series of technologically driven developments portends a radical transformation of the work of journalists. The three technological developments are the growth of social networking, mobile media and computational journalism. These developments are fueling a change in the basic paradigm of news reporting. As Thomas Kuhn (1962) noted in his seminal work, paradigm shift involves changing the basic assumptions that underline an entire discipline. The confluence of these three technological developments is enabling a shift from a highly controlled, often closed model of professional news reporting to a more open, collaborative, and engaged-citizen news reporting paradigm.

The advance of social networking, mobile and digital technologies is fundamentally challenging the basic assumptions that have guided professional journalists in their work as practiced in most of the developed world for at least a century. Perhaps most essentially, the assumptions are that: 1) professionally trained (either in journalism school, at a professional news organization or a combination of both) reporters gather the news for a commercial or non-profit news organization, 2) professionally trained editors at those news organizations act as gatekeepers to determine what news is published and in what form, and 3) news is formatted for distribution via centrally controlled media on a schedule determined by the owners and editors of those media.

Social networking and mobile, digital media have given rise to the wide spread role in news gathering by what have come to be known as citizen journalists. Preliminary evidence suggests the volume of citizen reported news is increasing (Goode 2009). Professor William Dutton (2009) of the Oxford Internet Institute suggests "networked individuals" is a better term for citizens acting as a fifth estate in the digital age. Of course, citizens acting as reporters have long played a role in newsgathering. Many point to the 1992 Rodney King beating video as a milestone in citizen reporting. But in fact, there have been many similarly profound moments in the history of citizen reporting. Perhaps the 8mm film of the 1963 assassination of U.S. President John F. Kennedy as recorded by bystander Abraham Zapruder (1963) is the most famous. But, the fact is that even in the Colonial era of the Americas much reporting was conducted lay citizens. But, for most of the 19th and 20th centuries, reporters in the employ of professional and usually commercial news organizations have done the majority of daily news reporting (Schmidt and Cohen 2013).

Citizen Journalism and the Rise of Social Networking and Mobile Media

Transforming the news gathering process in the second decade of the 21st century is the dramatic rise of citizens acting as independent observers using mobile digital devices and social networking technologies. This trend is particularly pronounced in the area of breaking news coverage, or reporting of news events as they unfold. Enabling this is the nearly ubiquitous reach of social media both domestically and internationally. There are some one billion registered users of Facebook and some five-hundred million registered users of Twitter, two-hundred million of them active Tweeters (Bullas 2012; Allfacebook 2012). Mobile media have grown to similar proportions, with at least one billion users of smart phones world-wide. There are more than one-hundred-sixty million tablet computer users in the U.S. (Pimentel 2013). These constitute a massive potential news gathering force worldwide. As a result, Schmidt and Cohen (2013) predict news media will exit the breaking news business.

With these extraordinary numbers, social media and mobile media have become highly influential in the media habits and behaviors of citizens in the U.S. and around the world, both as consumers and producers of news and information. The reliability of the information gathered is sometimes suspect, but the Internet social media space also provides a self-correcting function, as other critical users can point out errors. This trend is both the case for entertainment and news, as well as for social communication and for other communication purposes.

Social networking and mobile media are especially valuable for news media seeking to engage the younger segments of society, who have fallen away from the use of traditional news sources. They have most actively embraced social media and most often have disengaged from traditional media.

The Pew Research Center's Internet & American Life Project (Duggan and Brenner 2013) documents that youth are particularly apt to use social media. Pew reports, "Internet users under 50 are particularly likely to use a social networking site of any kind, and those 18 to 29 are the most likely of any demographic cohort to do so (83 percent). Women are more likely than men to be on these sites. Those living in urban settings are also significantly more likely than rural internet users to use social networking."

Fig. 1.: Citizen journalists William Gagan, right, and Geoffery Shivley, left, reporting from Syria in 2012 (Source: Photograph by William Gagan, used with permission)

Shown in Figure one are citizen journalists William Gagan, right, and Geoffery Shivley, left, reporting from Syria in 2012 (Devereaux 2012). These citizen journalists and others like them played a crucial role in news reporting during the so-called Arab Spring of 2011–2012. As noted in the *Guardian*'s report on citizen journalists in Syria, "Two Americans whose first foray into citizen journalism was filming at an Occupy protest have moved on to one of the most dangerous places in the world for reporters of any skill level." Professional journalists were unable to gain access to the country torn apart by civil war. "Yet without the backing of a media company and with little experience as journalists, two Americans, William Gagan, 30, and Geoffery Shivley, 29, decided to enter Syria in a risky attempt to film inside the divided country." Their reporting provided rare and much-needed critical information about developments in the country.

During the Arab Spring, reporters from professional news media often were unable to witness events themselves, as they sometimes unfolded rapidly or were in locations off limits to outside journalists, or at least very hard to access. Consequently, the only reporting, often in the form of photos and videos, was that captured by citizens at the events, often participating in protests, and shared via social networking sites. Sometimes, citizen reporting provided a complement to that produced by professional reporters. Yet, such citizen journalism is often difficult to verify. Citizen reporting may be subject to manipulation and taken in isolation may not be trustworthy.

Still, citizen journalism and its widespread growth presents a powerful challenge to the traditional role of journalists. It brings potentially dramatic changes to the work practices of journalists in the 21st century. Citizen journalists have embraced digital technology, especially mobile media, at an extraordinary pace and in an innovative fashion, in many ways far more so than traditional media companies. Traditional journalists historically have played a vital role in the democratic process by reporting information on matters of public importance to the citizenry, seeking to revitalize their relationships with the public.

Consider the case of the citizen journalism site, *AllVoices*. It has more than 375,000 registered citizen journalists and featured more than 1,000,000 reports between March 2012 and March 2013. These reports are mostly brief items shared on social media, including tweets, SMS messages, blog posts, photos and videos. Another example is *Mereporter*, a location-based, mobile app for citizen journalism, for iPhone and Android devices. It illustrates the new class or mobile tools available for journalistic work today.

The role of social media, mobile media and digital sensors (e.g. cell phone records, security cameras, facial recognition software) in news gathering came to a climax in March 2013 in Boston, MA, when a bomb attack was executed at the end of the annual Boston Marathon. The confluence of data produced by these sources not only contributed to law enforcement's identification and pursuit of two suspects in the case, but also to a non-stop news media feeding frenzy that ensued during the manhunt or suspect pursuit. Photos and videos captured by citizens and security cameras quickly became sensational evidence widely distributed online and across all news media in the U.S. and internationally. Millions followed the story online, as they are increasingly doing around the world. Zassourky (2010) provides evidence that the public in Russia and elsewhere is increasingly turning to social media for breaking news because of their speed and ubiquity. Law enforcement agencies such as the Federal Bureau of Investigation (FBI) used Twitter to communicate with not only the press but the public. Citizens and reporters alike used a free app to listen to the police scanner in real-time, following events as they unfolded in the Boston suburb of Watertown, MA, where law enforcement eventually engaged the suspects in a wild gun battle, killing one suspect and capturing the second. Video of the shoot-out was captured by a citizen observing from her home, and even offering her own real-time commentary much as a professional reporter might. This video was shown widely by mainstream news media.

Mobile technologies along with other changes (cost-cutting and staff reductions) are simultaneously influencing news reporting in sometimes less-than-beneficial fashion, by stimulating often superficial or producing erroneous reports delivered in breath-taking dispatches on Twitter or elsewhere where depth, thoughtful reflection or fact-checking may not occur. This problem is exacerbated by the significant cuts in editorial staff, especially among the most experienced journalists because of their higher salaries, at many news organizations, particularly newspapers. Ironically, while the advent of citizen reporting is adding to the mix of news feeds, it is also contributing to an acceleration of news reporting and an apparent increase in reporting errors. These errors are not necessarily by the citizen reporters.

Oftentimes, as was the case in the 2013 Boston bombing case, professional news reporters and commentators anxious to contribute to the flow of breaking news committed the errors. The Associated Press, CNN, and *The New York Post* all reported erroneously during the breaking news event. Subsequently, media critics pointed to the mistakes, especially an erroneous premature report that law enforcement had arrested a suspect in the bomb attack (Weber 2013). Law enforcement harshly criticized this reporting error, as possibly adversely affecting the FBI investigation. CNN and others had reported this error attributed to an unnamed source and without fact-checking for accuracy from at least one other source. Even when not reporting incorrectly, much of the reporting has been reduced to breathless real-time banality. Such was the case at CNN as an on-air reporter anxiously reported as several cars drove by: "Something just happened [...] we don't know what it was [...] that's the third canine dog [...] interesting that dog is barking [...] it may have smelled what we smelled."

Computational Journalism

The convergence of computer science and journalism, or computational journalism (CJ), is advancing a wide variety of new capabilities for news gathering and storytelling. Among the most significant is the introduction of artificial intelligence into the journalism process. This is apparent in the increasing use of algorithms by news media. An algorithm refers to a specific set of instructions to carry out a procedure or to solve a problem, typically in the form of a computer program (Wolfram Mathworld 2013). In journalism, algorithms can be used to complete a variety of tasks, especially related to the interrogation of large data sets including those compiled online. Using algorithms can help reduce costs for the news organization. Perhaps more importantly, it challenges journalists to reconsider their role and work methods. Journalists can use algorithms in a variety of ways. They can use algorithms to document their work, protect IP, create new types of stories, and identify and communicate with sources, citizen journalists, editors and colleagues also in the field and possibly distributed over wide geographic distances. Journalists have for many years used social networking sites such as Twitter, Facebook and YouTube to gather news, identify sources, and report on both breaking news and enterprise stories. In concert with algorithm data-mining, these tools are especially useful in quickly and efficiently identifying eye-witnesses to news events.

Diakopolous (2010; 2012), Naaman, and de Choudhury (2011) have conducted research on the role of CJ and social media for rapid source identifcation. They have developed computerized tools, or algorithms, for analyzing Twitter feeds for their utility in news reporting. In one case Diakopolous et al. used a "Seriously Rapid Source Review" (SRSR) computer-based tool to identify from Twitter feeds and other social and mobile media some of the eye-witnesses to the Tottenham Riots, which occurred in a neighborhood in London, England on August 7, 2011. Eye-witnesses are vital to reporters as sources of information for interviews in developing stories on breaking news such as the riots. Identifying eye-witnesses rapidly and efficiently has long been a challenge for journalists.

Implications for News Content

Journalists use algorithms to mine data sets and produce news visualizations and news reports. From *The Guardian* to *The New York Times*, a growing number of news organizations are using algorithms to tap into Big Data and build data-driven visualizations, whether reproduced in print or online in interactive and three-dimensional format. News media can also use algorithms to write news stories […] and possibly replace or supplant human reporters. Pew reports in its State of the News Media 2013 study, "a growing list of media outlets, such as *Forbes* magazine, use technology from a company called Narrative Science to produce content by way of algorithm, no human reporting necessary." This is both a cost-savings strategy and a move toward more rapid, efficient, and potentially customizable reporting. It has been reported as many as 30 clients from the Big Ten Sports Network to publisher Hanley Wood are using the software to produce stories.

Following are two versions of a news *lede* (i.e. sometimes spelled "lead," it refers to the beginning of a news story that summarizes the main points of the story), one reported and written by a computer algorithm, called Quill, a branch of artificial intelligence, and the other by a human reporter.

1. "The University of Michigan baseball team used a four-run fifth inning to salvage the final game in its three-game weekend series with Iowa, winning 7–5 on Saturday afternoon (April 24) at the Wilpon Baseball Complex, home of historic Ray Fisher Stadium."
2. "Michigan held off Iowa for a 7–5 win on Saturday. The Hawkeyes (16–21) were unable to overcome a four-run sixth inning deficit. The Hawkeyes clawed back in the eighth inning, putting up one run."

Version one is a relatively long lede (43 words) by journalistic standards, with the author attempting to pack in as much information as possible. The lede answers several of journalism's so-called "Five Ws," or who (Michigan baseball team), what (won the final game in a series), when (over the weekend, Saturday), where (Ray Fisher Stadium), and why or how (with a four-run fifth). But, forty-three words is far too long for a well-written lead. The standard in professional journalism is no more than thirty words […] beyond that the lede gets unreadable. Long ledes are often the product of either poor editing or unskilled reporters who lack the experience to judge which elements are most important and how to structure a lead. Version 1 is human written. Version two, the superior version by most professional journalism standards, was reported and written by the Narrative Science algorithm, Quill, co-developed by Larry Birnbaum, a professor at Northwestern University with a Ph.D. in computer science (Birnbaum 2013).

Narrative Science also uses its algorithm to automatically write stories using Twitter feeds. The company's stated goal is to use algorithms to produce stories for an audience of one […] the ultimate in customized storytelling. Columbia University's Newsblaster is another algorithm-based tool for automatically generating news stories. Quill wrote 370,000 stories in 2012 and is expected to write more than two million in 2013, most of them on demand for very small, highly targeted audiences. Algorithm-driven news reporting and writing is already transforming a variety of journalism topical domains. This is especially the case where formulaic approaches are common. Sports, business and finance, medicine and technology are some of the fields where products such as Narrative Science are already making their mark. Human journalists face a challenge to adapt and specialize in activities that computers cannot, at least at this time. Human reporters will need to focus on rigorous, enterprise and investigative reporting. Journalism educator Peter Laufer (2013) admonishes journalists to slow down. He recommends they take the time to carefully gather all the facts and thoughtfully report them. Laufer explains: "We must eat in order to survive. Accurate information can be another requirement for our survival. Yet our quest for instant information has made it more difficult to find the truth and see the larger picture behind breaking events." Human writing that focuses more on the why and the context that underlies a story will provide value. Examining why and writing with depth, nuance and rigorous reporting can be the hallmarks of human journalism. As Laufer observes: "We need to ask ourselves what news is important and why." These questions should drive the human journalist.

Mining of geographically tagged Twitter posts also present important implications for news content. Reporters can use geo-tagged tweets to track news and news-related discussion. They can use such geo-tagged tweets to identify topics or problems of interest to different communities, at least as reflected in the topics people are tweeting about. In one project in Qatar, researchers at the Qatar Computing Research Institute (QCRI) have tracked tens of thousands of tweeters in terms of the location of their posts, the topics tweeted about, and the language in which they tweet (Weber 2013).

Robotics is a second branch of AI with significant implications for journalism, especially news gathering. Robotics is fueling explosive growth of commercial drones. Once thought of as exclusively a military technology, these small, human-piloted aircraft are equipped with cameras and wireless transmitters and are poised to become widespread news gathering tools from the sky. The cost of drones is falling rapidly, and is now less than one-thousand dollars, which is less than the cost per hour for a human-piloted news helicopter. And the skill needed to remotely fly a drone is substantially less than for a manned helicopter. Drones also reduce news media reliance on government data sources. They enable journalists to gather and report using data rather than sound-bites. This is especially useful in environmental, traffic and weather reporting. Drones are also much safer than traditional air-based news gathering. Since 1979, there have been seventy-one crashes of news gathering helicopters with ninety-one deaths world-wide (Smith-Spark, Laura and Nick Thompson 2013). Yet, ethical questions regarding the appropriate use of drones are significant, particularly in the context of privacy and security. Legislators in a number of communities across the U.S. and foreign governments are considering laws to prohibit or severely curtail the domestic use of drones by news media or anyone else (Herzog 2013).

Also important to journalistic work are new mobile devices such as the Samsung Galaxy S4 smartphone that incorporate AI capabilities such as gesture and gaze control. These new capabilities can be used in news gathering and storytelling. Dual-facing cameras allow two photos taken simultaneously in two directions to be seamed automatically into a single photo. This technology enables, for instance, a viewer to record watching a game and simultaneously getting fan reaction. The Samsung S4 also features gesture and gaze control of the smartphone, giving a reporter hands-free control of the device. This could be useful during breaking news coverage when simplified device control could facilitate note taking, interviewing or direct observation of an event.

Consider also the potential health implications of AI-enabled digital technology for journalists. *New York Times* reporter Lisa Foderaro reportedly suffers from repetitive stress injury, or RSI, as a consequence of extensive writing and other use of a computer (Monua 1994). She now uses computer speech recognition to write her stories. As a branch of AI known as "natural language," speech recognition has advanced dramatically in recent years. One mobile application called Dragon Dictation illustrates. Earlier this spring a faculty member told the author he was having trouble using a computer because of repetitive stress injury, or RSI. The author encouraged the faculty member to consider using computer dictation software using speech recognition. Two decades ago such technology was only about 80 percent reliable […] as a test, the author delivered an entire semester's worth of lectures using dictation software to create a real-time transcript that students could use, especially those with a hearing disability. In this semester's case, the author suggested the Dragon dictation app, which is free for iPad, as the faculty member in question said he relied primarily on the iPad for his work. Following are the results of a quick test the author conducted speaking into the Dragon app:

> "Hello, this is John Pavlik of Northwestern University in Qatar. I'm testing Dragon naturally speaking dictation software for iPad. I'm evaluating its reliability as a dictation tool."

There are 183 characters in this passage, and the Dragon app made no errors. And here's the faculty member's response to the author's Dragon suggestion:

"Dear John this whole e-mail has been written through Dragon I hope it works well I hope you will see all that I have to say so far this is really cool thank you very much for the recommendation."

The only dictation skill he needed to master is adding punctuation … but it is an easily learned skill.

In fact, here is the epilog to the story. A few minutes later he sent me this email, also created by his spoken word:

"I just figured out how to punctuate and was going to email you. Thanks again for this. It really does help. With appreciation."

Developments in citizen journalism and digital technology have also led to significant changes in the content of news at traditional news organizations and further changes are on the horizon. News organizations have adapted in many ways to this changing environment by incorporating citizen journalism, or user-generated content (UGC) in their news reports. This is leading to an increased emphasis on the curation, verification and gatewatching roles of professional journalists. Curation refers to journalists sifting through UGC to identify content considered news worthy and then re-presenting it via traditional media channels, often in edited form. Verification refers to the process of independently confirming the veracity, accuracy or reliability of the news and information reported in UGC. Gatewatching is a new journalistic function Bruns (2005) identified to describe a transformation from the traditional news media role of gatekeeping. In the past, news media played a key role in setting the agenda for matters of public importance by their selection of news stories and their presentation. In the digital age, news media are increasingly acting as intermediaries in the flow of news and information presented in social media, filtering and reorganizing material to pass along to news consumers. Research suggests these functions are increasingly rapidly in 21st century professional news media.

CNN has created the CNN iReport as a mechanism for incorporating citizen journalists' video reports. iReport has provided unique video reporting on a wide swath of stories ranging from tornadoes in the Midwest to political reports on the presidential candidates in 2012 (iReport). Illustrative is a report on a controversial killing under Florida's stand-your-ground law. As told on the iReport, "New York photographer Joel Graham stood in the crowd on Wednesday when hundreds gathered to demand justice for the killing of Trayvon Martin, the unarmed teenager shot by a neighborhood watch volunteer in Central Florida last month."

Researchers such as Silva (2011) at the University of Texas-Austin are studying the impact of the CNN iReport and other citizen journalism on the practice of journalism more generally. Her research suggests media are increasingly incorporating UGC into mainstream news media channels. Mainstream media are relying on UGC to provide more nuance to overall reportage, especially providing reporting capability when professional resources are limited. An example comes in the form of UGC offered on the CNN iReport on the anniversary of the Occupy Wall Street protests. As noted on the iReport, "Photographer Shameel Arafin has been covering the Occupy Wall Street protests since they began last fall. He was at the six-month anniversary rally on Saturday, when protesters tried to re-occupy Zuccotti Park."

Location-aware mobile technologies are also enabling new types of news content. Journalists can use Global Positioning Satellite (GPS)-enabled mobile media to create new forms of geo-located content, including augmented reality (AR)-enhanced content. Such AR-enhancements can be embedded into print and other media or geo-tagged to real-world locations. A growing number of news organizations are experimenting with AR-enhanced storytelling. News photos, for instance, can be tagged with related videos or animations. The AR-enhancements are accessible by the news consumer using an AR-app on a mobile device pointed at the photo. Among the news media to publish reports using AR-enhancements are *USA Today*, *The New York Times*, *The Boston Globe* and the *Süddeutsche Zeitung*.

This may give rise to a new form of journalism, what might be called "Geo-located Journalism," where stories are geo-graphically tagged and embedded throughout both the physical and virtual world via Google Earth. Geo-located Journalism requires new methods and especially new ways of thinking about storytelling and the audience, now more a participant in journalism than a consumer.

Restructuring the Newsroom

Social networking and mobile media are also influencing newsroom structure. One of the forms of journalism that has grown in the 21st century as digital and mobile technologies have supported new organizational structures is Hyperlocal news. AOL's Patch.com is a model in the hyperlocal arena. It offers customizable, mobile, geo-located news. Veteran journalist John

Hall's Redbankgreen, NJ was one of first hyperlocal news sites, and operates independently.

Patch tripled its audience traffic between 2010 and 2012, with almost ten (9.9) million unique visitors per week. Yet, with its one-thousand local news sites in two dozen states, Patch has also reportedly lost an estimated one-hundred million U.S. Dollars so its business model has not yet turned the financial corner. Its sustainability is also in some doubt (McMillan 2012). Patch employs more than one-thousand professional journalists to manage its hyperlocal sites, and act as editor for more than thirteen thousand citizen journalists contributing content. Hyperlocal journalism is a model that may extend to Europe, Asia, Latin American and Africa, particularly in the structure suggested by Patch with experienced, professional journalists acting as site managers and citizen journalists or networked individuals using mobile devices and social media to contribute original content for their local communities. They need training though in reporting techniques and values, as well as storytelling methods. Professional journalists need to act as mentors.

The convergence of mobile media, data journalism and computational journalism is also enabling the creation of highly efficient virtual news organizations that can produce high quality news at relatively low cost. An example is Inside Climate News. The reported budget for Inside Climate News is five-hundred and fifty thousand U.S. Dollars, most of which pays for staff. It also pays for the travel and Internet services (Stelter 2013). It does not need to pay for rent or other costs associated with the physical infrastructure long associated with 20th century journalism. For more than a century, newspaper barons from William Randolph Hearst to Rupert Murdoch have built edifices to both house their employees and requisite equipment (e.g. printing presses) as well as to reinforce their public credibility. In an era of near media monopolies enjoyed by many news media in most communities, as well as high barriers to entry to publish or broadcast the news, such physical extravagance was little more than an affordable luxury. But in an age of decimated revenues, falling audiences and heightened competition for commoditized digital news, such expenditures can no longer be rationalized or justified financially.

In contrast, the highly lean and low-cost Inside Climate News organization founded just five years ago won the 2013 Pulitzer Prize for national reporting. It beat other far more established news media, including *The Washington Post* and *The Boston Globe*. Inside Climate News's coverage of a major oil spill in Michigan featured a seven-month investigation by a news staff of three geographically distributed reporters working collaboratively online. *The New York Times* said: "'We're a virtual organization,' said the publisher of the six-year-old Web site, David Sassoon, from his office in New York. So the celebration took place in a telephone conference call; whatever Champagne flowed, flowed in separate locations.'" The reporters who worked on the story were each in a different city. "Elizabeth McGowan was in Washington, Lisa Song was in Boston and David Hasemyer was in New York." Funding the non-profit Inside Climate News is a revenue model built on individual donations from the public and charitable foundations.

The Project for Excellence in Journalism 2012 State of the Media report states that technology intermediaries (e.g. Google, Apple) have taken control of the news. Digital technologies present unique opportunities to restructure and thereby build stronger relationships with the citizenry. Through these tools journalists can re-establish the once vital role played by local, regional and national journalism in the democratic process in communities both domestically and internationally. An intriguing example comes in the form of the online citizen initiative from India called ipadabribe.com (2014).

Created by the *Janaagraha Centre for Citizenship and Democracy* in Bangalore, India, the site encourages citizens to report on bribes they have been forced to pay to obtain government or other services. The site is a form of citizen journalism. It is also Big Data. As of this writing, February 4, 2014, it has tallied some 24,368 bribery reports in 635 cities from around India (up to 58 percent since March 2012, when 15,449 reports had been tallied). Stories include citizens forced to pay bribes to the police for a variety of routine services, to government agents to obtain passports, or to a customs agent to retrieve a laptop computer forgotten at the airport.

Some traditional news media are collaborating with ipaidabribe.com to create and disseminate news stories based on the citizen journalism reported on the site. *The Times of India* reported: "Bangaloreans paid a whopping Rs 11.35 crore" (11.35 million, or in US$ 210,000) as a bribe on 4,339 occasions. "A major chunk of money went for 'managing' parking violations."

Cyber-Security and Journalism

One issue of critical importance in an age of online journalism is how to defend against cyber-attacks on news media around the world. Cyber-security is a critical issue for news media and journalists should employ encryption and other digital tools as a defense against hackers or governmental surveillance (e.g. NSA's PRISM) (The Unhived Mind 2013). As freelance and full-time media employed journalists increasingly rely upon mobile, wireless media in news gathering, production, distribution and communication, they are vulnerable to government and others who would use digital, networked technology to monitor, control or censor their work. A case in point, although not journalism, comes from 2009. In July of that year Amazon remotely and clandestinely accessed via the public telecommunications network the Kindle e-reader devices of persons in the U.S. who had bought e-book copies of George Orwell's classic *1984*. Overnight, Amazon deleted those copies, without telling anyone or asking their permission. In the morning, Kindle owners seeking to read *1984* no longer found it on their devices, and eventually Amazon admitted what it had done. Its explanation was that the company learned that it did not have the legal right, the copyright permission, to distribute the e-book in the U.S. Company executives had decided deleting the e-copies was the best way to proceed. Many Kindle owners did not agree and there was an uproar, and eventually an apology from Amazon, followed by a lawsuit.

The implications for journalists are clear: Whatever is on a digital device can be spied upon, hacked, deleted […] journalists can digitally protect their work, both to guard their intellectual property (IP) rights but also prevent hackers from maliciously sending out false messages. Such was the case on April 23, 2013, when Syrian hackers broke into an unsecured Twitter account of the Associated Press and at 10:07 am Eastern Standard Time (EST) in the U.S. sent out a false tweet alleging explosions at the White House had injured President Obama (Welch 2013). Within minutes, stocks on Wall Street plunged an estimated two billion dollars in value, partly due to computer trading triggered by key words such as "Obama injured," "White House" and "explosion." Eventually, the financial losses were recovered but the point had been made: False Tweets (and other social media) have the potential to do real harm.

Geo-tagging is useful for the protection and authentication of content. Digitally tagging, or time, date and location stamping, photos and video such as that from Kinomap, can enable the independent verification of where and when such content was produced and by whom. This can be valuable in the battle against plagiarism and hacking. It can help the entrepreneurial journalist protect her or his IP.

Concluding Reflections: An Emerging Paradigm of Open News Reporting

Developments in technology have long spurred changes in the work of journalists. In recent years, the confluence of social networking, mobile media and digital technologies has fueled a transformation of the traditional model of newsgathering, particularly during breaking news coverage. Citizen journalists are increasingly complementing or replacing professionally trained journalists in the reporting of breaking news, particularly via social and mobile media. Data-driven stories are playing an increasing role in journalism, supplanting a long-standing approach emphasizing personal interviews. A convergence of computer science and journalism is advancing the use of algorithms and other forms of artificial intelligence in both news reporting and storytelling. Control is shifting from traditional media gatekeepers into a collaborative mixture of professional journalists and lay citizens equipped with mobile devices connected via a global social network and built on Big Data.

Figure two summarizes the characteristics of this emerging open, data-driven reporting paradigm in terms of the core qualities and practices of journalism. These characteristics represent testable hypotheses (e.g. an increasing percentage of breaking news is being reported by citizen journalists using social and mobile media) and a potential research agenda for the 21st century.

The consequences of this shift in news reporting are potentially far reaching. In this emerging paradigm, innovative journalists can draw upon the principles of data mining, citizen journalism and digital technologies to create a new form of aggressive, collaborative and independent public service journalism to invigorate the fourth, and *fifth*, estates in the 21st century.

News Qualities	*Controlled Paradigm*	*Open Paradigm*
Fact checking	Pre-publication	Crowd sourced
Speed	Slow, scheduled release	Real-time
Storytelling	Inverted pyramid, linear narrative, factual	Dynamic, interactive storytelling, data-driven, algorithm influenced, non-linear narrative, factual
Customization	Limited, geographic	Extreme, algorithm driven
Multimedia	Limited, planned	Extreme, spontaneous, ubiquitous, citizen and sensor driven
Ethics	Professionally defined codes	Collaboratively defined
Newsroom Structure	Centralized, hierarchical	Virtual, flattened, dynamic
Sources	Interviews (documented, controlled), identified, anonymous and confidential, limited first-hand observation and data interrogation	Transparent, documented interviews, extensive data-driven interrogation, first-hand observation

Fig. 2: Characteristics of an emerging open reporting paradigm (Source: author's illustration)

Works Cited

Allfacebook (2012). 03/05/2013, http://www.allfacebook.com/facebook-users-2012–02.

Birnbaum, Larry (2013). *Intelligent Information Laboratory*, 02/04/2014, http://info-lab.northwestern.edu/people/larry-birnbaum/.

Brownlow, Mark (2012). *Smartphone Statistics and Market Share. Email Marketing Reports*, 03/05/2013, http://www.email-marketing-reports.com/wireless-mobile/smartphone-statistics.htm#smartphones.

Bruns, Axel (2005). *Gatewatching: Collaborative Online News Production.* New York: Peter Lang.

Bullas, Jeff (2012). *Twitter Approaching*, 03/05/2013, http://www.jeffbullas.com/2012/02/13/twitter-approaching-500-million-users/.

Devereaux, Ryan (2012). Citizen Journalists Undertake Mission in Syria: 'Bullets Don't Discriminate.' *The Guardian*, March 5, 02/04/2014, http://www.theguardian.com/world/us-news-blog/2012/mar/05/american-citizen-journalists-syria.

Diakopoulos, Nick (2010). *A Functional Roadmap for Computational Journalism*, 03/05/2013, http://www.nickdiakopoulos.com/2011/04/22/a-functional-roadmap-for-innovation-in-computational-journalism/.

Diakopoulos, Nick (2011). *Blog*, 03/17/2012, http://www.nickdiakopoulos.com/blog/.

Diakopoulos, Nick (2012). Munmun De Choudhury, and Mor Naaman. Finding and Assessing Social Media Information Sources in the Context of Journalism. *Proceedings, CHI 2012, May 2012*, 03/18/2012, http://sm.rutgers.edu/pubs/diakopoulos-srsr-chi2012.pdf.

Duggan, Maeve, and Joanna Brenner (2012). *The Demographics of Social Media Users 2012*. Pew Research Center's Internet & American Life Project, 02/04/2014, http://www.pewinternet.org/Reports/2013/Social-media-users.aspx.

Dutton, William H. (2009). The Fifth Estate Emerging through the Network of Networks. *Prometheus*, 27, 1–15.

Goode, Luke (2009). Social News, Citizen Journalism and Democracy. *New Media & Society*, 11, 1287, 05/02/2013, http://ruraleconomics.fib.ugm.ac.id/wp-content/uploads/Luke-Goode-Social-News-Citizen-Journalism-and-Democracy.pdf.

Herzog, Blake (2013). California City to Vote on Banning Drones. *USA Today*, April 4, 02/04/2014, http://www.usatoday.com/story/news/nation/2013/04/04/rancho-mirage-hobby-drones-ban/2052193/; http://www.cnn.com/2013/03/20/us/drones-domestic-use; http://ip/aidabribe.com/.

iReport (2013), 03/28/2012, http://www.cnn.com/ireport/archive/index.html.

King, Rodney (1992). Video of Rodney King's Plea during the 1992 Los Angeles Riots, *YouTube*, 02/04/2014.

Kuhn, Thomas S. (1962). *The Structure of Scientific Revolutions*. Chicago: University of Chicago Press.

Laufer, Peter (2013). *Calls for Slow News Movement*, 01/16/2013, http://www.slownewsmovement.com/2013/01/16/peter-laufer-calls-for-slow-news-revolution/.

McMillan, Graeme (2012). *AOL's Hyperlocal 'Patch' Network Triples Traffic in 2011, But What Does That Mean?*, 03/22/2012, http://techland.time.com/2012/01/11/aols-hyperlocal-patch-network-triples-traffic-in-2011-but-what-does-that-mean/.

Monua, Janah (1994). *Newstalk at the New York Times. Highbeam Business*, 04/03/2014, http://business.highbeam.com/392705/article-1G1–15894294/newstalk-new-york-times.

Morse, Samuel F. B. (1839). *New York Observer*, April 20. In Flickr (2013), *The History of Photography*, 02/04/2014, http://www.flickr.com/photos/photohistorytimeline/4037283778/in/photostream/.

Pimentel, Benjamin (2013). Tablets surge but Apple loses market share. *Marketwatch*, May 1, 02/04/2014, http://articles.marketwatch.com/2013–05–01/industries/38950539_1_tablet-shipments-market-share-ipad-mini.

Pool, Ithiel de Sola (1983). *Technologies of Freedom*. Boston, Mass.: Belknap Press.

Schmidt, Eric, and Jared Cohen (2013). *The New Digital Age*. London: John Murray Publishers.

Silva, Cláudia (2011). *User Generated Content and Participatory Culture: How Has It Taking Place in CNN's Website?* Congress Paper, SBPJor 'Journalism and Digital Media,' Rio de Janeiro.

Smith-Spark, Laura, and Nick Thompson (2013). *2 Killed as Helicopter Crashes into Crane in Central London*, January 16, 03/05/2014, http://www.cnn.com/2013/01/16/world/europe/uk-helicopter-crash.

Stelter, Brian (2013). Inside Climate News Hopes to Build on Pulitzer. *The New York Times*, April 17, 03/04/2014, http://www.nytimes.com/2013/04/17/business/media/insideclimate-news-hopes-to-build-on-pulitzer.html?_r=0.

The Unhived Mind (2013). *Encryption Communication Services Gaining Momentum after NSA Prism Spying*, 06/30/2013, http://theunhivedmind.com/wordpress3/2013/06/25/encryption-communication-services-gaining-momentum-after-nsa-prism-spying/.

Toumi, Habib (2013). *Kuwait Blogger Jailed for 2 Years*, 02/04/2014, http://gulfnews.com/news/gulf/kuwait/kuwait-blogger-jailed-for-2-years-1.1162110.

Weber, Peter (2013). *Watch: Jon Stewart Slams CNN's Boston Bombing Coverage*, 02/04/2014, http://theweek.com/article/index/243115/watch-jon-stewart-slams-cnns-boston-bombing-coverage.

Welch, Chris (2013). *AP Twitter account hacked*, 02/04/2014, http://www.theverge.com/2013/4/23/4257392/ap-twitter-hacked-claims-explosions-white-house-president-injured.

Whitman, Walt (1846). The Brooklyn Eagle. In The Library of Congress, *America's First Look into the Camera: Daguerreotype Portraits and Views, 1839–1862*, 02/04/2014, http://www.loc.gov/teachers/classroommaterials/connections/daguerreotype/langarts.html.

Wolfram Mathworld (2013). *Algorithm*, 02/04/2014, http://mathworld.wolfram.com/Algorithm.html.

Zapruder, Abraham (1963). *Film of Assassination of President John F. Kennedy*, 02/04/2014, http://en.wikipedia.org/wiki/Abraham_Zapruder.

Zassoursky, Ivan (2010). *Social Media and Reporting the News in Moscow*. Conference Paper, Digital Media Conference at Moscow State University, Moscow.

Journalistic Quality as Crowd Wisdom? What Journalists Think about Criticism on the Social Web

Tobias Eberwein

Introduction: Of Toothless Tigers and Cyber Watchdogs

Amidst the cataclysms that are currently reshaping the international media landscape, there is much evidence that the established system of media self-regulation and journalistic quality management is facing a severe crisis: For decades, press and media councils all over the world have had to contend against the criticism that they are "toothless tigers" without any impact on practical newsroom work (e.g. Pöttker 2003). Today, the ongoing discussions about consequences from the recent *News of the World* scandal have done their share to discredit institutions like the British Press Complaints Commission even further—not only in the United Kingdom, but in many European countries, where debates about a reform of the principles and processes of media self-regulation have reached a high degree of controversy (e.g. Jempson 2013). Alternative instruments of journalistic self-observation are no big help in the current situation either: The concept of the ombudsman, for example, is still underdeveloped in many journalism cultures worldwide (Elia 2004)—and if an ombudsman position exists at a media outlet, it is often seen as a financial liability rather than a source of journalistic quality, which can easily be eliminated without doing much damage (Starck 2010). Similar considerations are at play when it comes to judging the benefits of media journalism—in the sense of journalistic reporting about journalism and the media: In theory, it may contribute to the self-reflection of journalistic actors, but in practice, it is regarded as a dispensable asset which finds itself on the top of many publishers' lists of possible cost reductions because they assume that the readers are not interested in this kind of navel-gazing anyway (e.g. Beuthner and Weichert 2005).

But how can we make sure to maintain a justifiable standard of quality in journalism when the traditional instruments that are supposed to safeguard this standard fail to do so? Certainly, intensified regulatory measures by state actors are not an option, if we take the idea of press freedom seriously.

Over the past few years, however, various observers have noted that another stakeholder group with an intrinsic interest in a free and pluralistic media landscape might offer valuable assistance: the audience, which is currently being (re-)discovered as one of the favorite objects of analysis in the realm of communication and media studies (Loosen and Schmidt 2012). Backed by the technological changes of the recent past, particularly the development of the Internet, members of this group have found new ways for expressing their thoughts about the media's performance: They directly comment on journalistic products by using the comment function that has become a standard feature of many online news websites (e.g. Springer 2011; Ziegele et al. 2013); they maintain watch blogs that are designed to keep track of journalists' mistakes (e.g. Eberwein 2010; Fengler 2008); they use social networking platforms like Twitter or Facebook to communicate about their experiences as TV viewers (e.g. Buschow et al. 2013; Wohn and Na 2011). With these web-based instruments of media observation, ordinary media users turn into kinds of cyber watchdogs that fulfill functions similar to the traditional institutions of media self-regulation: They kick off a public discussion about the quality of journalism and the media which is supposed to complement or even replace formal mechanisms of controlling the so-called "fourth estate." However, unlike more traditional forms of media watching and facilitated by technological innovations, they rely on the "wisdom of crowds" (Surowiecki 2005) to generate a judgment about what distinguishes good journalism from bad, thus giving rise to a novel kind of participatory media regulation that was largely unknown in the pre-Internet era.

But how effective and how desirable are these new participatory instruments of media observation? Can they really counterbalance the deficits of traditional media self-regulation and journalistic quality management? How do journalistic actors perceive their new cyber watchdogs? And how do journalists' attitudes towards media self-regulation and accountability differ across journalism cultures and media systems? These and similar questions led to the formation of a multinational research consortium which conceptualized and implemented the comparative research project "Media Accountability and Transparency in Europe" (MediaAcT).[1] The following sections of this paper discuss some of the results and considerations that orig-

1 Further information on the concept and the findings of the EU-funded MediaAcT project are available on the project website (http://www.mediaact.eu, 02/20/14)—and in two recent book publications (Eberwein et al. 2011; Fengler et al. 2014) which serve as a starting point for the research presented in this paper.

inated from the MediaAcT study. Special attention is paid to the project's comparative survey among journalists from 14 European and Arab countries which constitutes a meaningful empirical basis for answering the aforementioned key questions.

Background: Participatory Media Regulation—Past, Present, and Future

During the last decades, most Western democracies have developed a complex legal framework that is supposed to safeguard a free and pluralistic media landscape (Psychogiopoulou 2012). However, while media laws usually regulate the structures of media practices in the respective areas of application, journalistic contents are protected almost entirely from state interference (Puppis 2009, 57, 61). Therefore, it is even more important that media treat their various stakeholders responsibly, so that they can fulfill their manifold social functions without external control (McQuail 2013).

Media researchers have used a most diverse terminology to describe the processes which journalistic actors trigger in order to ensure responsible behavior. The terms media self-control (e.g. Baum et al. 2005; Stapf 2006) or media self-regulation (e.g. Puppis 2009) are commonly used to denote those practices which members of the journalistic profession initiate themselves to guarantee the quality of their coverage. The broader concept of media accountability, on the other hand, discusses "any non-State means of making media responsible towards the public" (Bertrand 2000, 108)—and thus not only includes journalists, but also media users and other stakeholders of the media into the process of journalistic quality management. In recent years, the concept of media transparency (e.g. Meier and Reimer 2011) has gained increasing scholarly attention. It focuses on a variety of instruments, particularly on the level of the media organization, that can contribute to preserve or regain trust in journalism by providing information about newsroom processes and the participating actors (e.g. with the help of online profiles of journalists, public mission statements, links to original sources, newsroom blogs etc.).

The MediaAcT project has combined the perspectives of media accountability and transparency, because together they promise to provide the most comprehensive picture of the potentials and problems of journalistic quality management under the conditions of digital communication (e.g. Fengler 2012). In this approach, media accountability instruments (MAIs) are not only traditional institutions within the journalistic profession (such as press and media councils, codes of ethics, ombudspersons, or media journalism), but also external institutions like media research, NGOs, or viewer/user associations of any kind. In addition to these institutionalized MAIs, the MediaAcT consortium also analyzes various innovative instruments of media observation that are currently flourishing on the Internet, both inside and outside of the journalistic profession—e.g. journalist and citizen blogs, cyber-ombudspersons, or media criticism via Twitter and Facebook.

For this paper, however, it is expedient to maintain a differentiation between the concepts of media self-regulation and participatory media regulation (Eberwein, Leppik-Bork and Lönnendonker 2013), in order to evaluate the impact of varying media accountability processes, depending on their point of origin. When the term media self-regulation is used in the subsequent paragraphs, it is supposed to refer to any non-state instrument of media accountability that is initiated within the journalistic profession. At the same time, the term participatory media regulation is used to label any MAI that is prompted by members of the audience. It is the central aim of this paper to compare the impact of these two groups of instruments of media accountability with the help of a systematic empirical research design.

Up to now, academic research has failed to provide reliable findings that could be used for such a comparison. However, recent studies that seem to prove the shortcomings of different traditional instruments of media self-regulation do exist in considerable quantity: For example, Campbell (1999, 755) concludes after an analysis of various MAIs in the U.S. media sector that her findings "do not provide a great deal of support for the claimed advantages of self-regulation." European scholars come to similarly skeptical conclusions (e.g. Pöttker 2010). Qualitative studies with media journalists in the USA and European countries (e.g. Fengler 2003; Porlezza 2006) show that even journalists who cover media issues for quality media shy away from criticizing their colleagues and supervisors. Studies dealing with ombudsmen (Evers, Groenhart and van Groesen 2010) reveal similar self-imposed restrictions. The list of studies with comparable results could be expanded almost arbitrarily, but this would hide the general deficit that most of the existing analyses of traditional media self-regulation lack an appropriate empirical basis—even more so if looked at from a comparative perspective.

In contrast, research on participatory media regulation is still in the fledgling stages: In fact, numerous studies have been published recently that deal with the subject of user participation in media communication, many of them also stressing the fact that participation is by no means a phenomenon that only began to thrive in the era of the Internet (e.g. Engesser 2008; Schönhagen 1995). However, only few of them discuss the potential benefits of participation as a means of journalistic quality management. If so, most of the recent publications focus on the influence of media blogs on journalistic performance: Although there is still a lack of systematic comparative analyses, there are the first indications that some media watch blogs may in fact have a higher journalistic quality than traditional journalism, e.g. with regard to timeliness, variety, comprehensibility, entertainment, interactivity, and hypermediality (Hutter 2009). Other studies point out that media criticism in blogs can be a potent means of monitoring the content and form of journalistic coverage, especially in the area of tabloid journalism (Schönherr 2008), and that users of watch blogs may be motivated to reflect on criteria for good journalism (Mayer et al. 2008). Moreover, media blogs can apparently serve as a tool for reflecting editorial decisions and generating user feedback (Theis-Berglmair 2009; Wied and Schmidt 2008). However, a comparison with the United States shows that, for example, the German media blogosphere is still underdeveloped, particularly with regard to a lack of sustainable business models and possible schemes for self-regulation (Fengler 2008). This critical assessment is backed by a content analysis (Eberwein 2010) which demonstrates that a key feature of many German-language media blogs is a lack of continuity in their reporting. Whereas a large number of their posts directly relate to news gathered by mainstream media journalists, the variety of their topics is even more limited than that of the media pages in the daily press.

Beyond the potentials of media blogs, research on participatory media regulation is much less versatile: A few researchers have fathomed the use of Twitter as a vehicle for TV criticism (e.g. Buschow et al. 2013; Wohn and Na 2011). A group of researchers in the MediaAcT consortium has provided several qualitative case studies that investigated further examples of online MAIs, in order to find out whether they can correct some of the well-documented deficits of traditional media self-regulation (Heikkilä et al. 2012). However, systematic comparative analyses of the impact of web-based media accountability processes are still missing.

Consequently, many important questions remain unanswered: How effective are the various non-state means of making responsible as opposed to state-controlled instruments of media regulation such as laws? Are participatory media accountability processes on the Internet a serious alternative to conventional regulatory approaches? How does the situation differ in journalism cultures within Europe and beyond? And what are possible implications for journalistic practitioners and media policy-makers? These questions constitute the corner points for the empirical research presented in this paper.

Method: An International Journalists' Survey

With the help of a comparative survey among journalists in 14 different countries, the MediaAcT project wanted to find out to what extent journalistic attitudes towards media self-regulation and audience criticism resemble or differ from each other in varying media systems and journalism cultures (Eberwein et al. 2014). We not only examined Western media systems in line with the classical typology by Hallin and Mancini (2004), who differentiate between "liberal" (in our study: the United Kingdom), "democratic corporatist" (here: Finland, the Netherlands, Germany, Austria, Switzerland), and "polarized pluralist" (here: Italy, France, Spain) system types. Additionally, and in contrast to Hallin and Mancini, we also included three Central and Eastern European media systems (Estonia, Poland, Romania) in different phases of political transformation. Moreover, with Jordan and Tunisia, two states from the Arab world moved into the focus of our study, which have only limited experiences with press freedom and media self-regulation up to now. This selection reproduces a so-called most different systems design (Przeworski and Teune 1970) which aims to compare as diverse populations as possible in order to generate universally valid conclusions.

All in all, from May 2011 to March 2012, we questioned 1,762 journalists in the 14 partner countries from our research project about their perspectives on and assessments of media self-regulation and user participation. As a starting point, the project teams in each country of analysis designed a stratified national sample which—depending on the availability of reliable statistical data—should represent the given basic populations of journalistic actors as precisely as possible. The first step of the stratification took into account nine different media segments: daily newspapers, weekly newspapers, magazines, public service radio, private commercial radio, public service television, private commercial television, online news media, news agencies. A second step in the sampling process distinguished between five different hierarchical positions in the newsroom: editor-in-chief, leading editor, reporter, freelancer, and trainee. This procedure was meant to do justice to the heterogeneity of the basic populations of journalists in the countries studied.

In order to calculate the sample sizes for each of the participating countries, it was necessary to determine the total sample size first. For this, we evaluated the response behavior in the pre-tests and defined the maximally estimated standard error as 0.05. With these data, it was possible to calculate a minimum total sample size of 1,762 (Buttler and Fickel 2002, 151), which was distributed proportionally among the participating countries according to their respective basic populations. In order to be able to evaluate the data, the minimum individual sample size was fixed at 100; the biggest national sample sizes, in those countries with larger basic populations, were as high as 237.

To reach these target values, the local project teams investigated the contact data of randomly chosen journalists—mostly with the help of disposable address databases (like the Austrian *Journalisten-Index* or the German *Zimpel*)—and invited these journalists via e-mail to participate in the survey. This procedure was repeated as long as it took to recruit national samples that completely matched the respective quota criteria. Altogether, almost 8,000 contacts were necessary. The average response rate of the MediaAcT survey reached 23 percent.

The standardized survey was realized online with the help of the EFS Survey (Unipark) software. In addition to the relevant social statistical data, the participants were questioned about their individual experiences with (traditional and web-based) instruments of media self-regulation, their evaluation of the culture of criticism in journalistic newsrooms, the influence of media management and journalistic profession on processes of media accountability, and the societal relevance of the topic. The questionnaire, consisting of 25 sub-questions, reprocessed the results of a comprehensive desk study on the status quo of media accountability in Europe and the Arab world (Eberwein et al. 2011) as well as a qualitative pre-study for which 98 international experts were interviewed about the prevalence and possibilities of web-based media accountability processes (Heikkilä et al. 2012). Therewith, the survey by the MediaAcT consortium may be classified as the first comparative communicator study that has systematically and extensively collected data about the attitudes towards and the potential effectiveness of different instruments of media self-regulation and accountability.

Findings: What Journalists Think about Criticism on the Social Web

The data from the MediaAcT study show that journalists from all over Europe and the Arab world have a rather high regard of the concept of media accountability and responsibility. When we confronted them with the statement "Journalistic responsibility is a prerequisite for press freedom" and asked them to rate this statement on a scale from 1 ("I totally disagree") to 5 ("I fully agree"), the degree of approval was remarkably high in almost all of the 14 countries studied (Tab. 1). The total mean of 4.46 is evidence of a clear commitment to responsible journalism. The fact that the deviation of the country-specific means was not bigger than ±0.22 in any of the national sub-samples makes it clear how much our respondents conform in the assessment of this item across the most different journalism cultures.

At the same time, however, our data suggest that this assessment might as well be lip service. In any case, the statement "Journalists are concerned about criticism they get from their audiences" only received an average rating of 3.35. Obviously, the consideration of audience criticism—actually one of the key characteristics of participatory media regulation—is far less common in everyday newsroom work as the general affirmation of the concept of media responsibility might lead the critical observer to assume. What this means for the effectiveness of different participatory instruments of media criticism still needs to be debated.

In order to assess the potential impact of different media accountability processes in more detail, the MediaAcT survey included the question "Which of the following have most impact on journalists' behavior in [country]?," which was complemented by a long list of traditional organizations of media self-regulation as well as various web-based media accountability instruments. Like before, the survey participants were invited to align their evaluation with a five-point Likert scale—in this case with the poles 1= "No impact at all" and 5= "Very high impact" (Tab. 2).

	Means	*Finland*	*Netherlands*	*Germany*	*Austria*	*Switzerland*	*United Kingdom*	*France*	*Italy*	*Spain*	*Estonia*	*Poland*	*Romania*	*Jordan*	*Tunisia*
Journalistic responsibility is a prerequisite for press freedom.	4.46	0.05	-0.22	-0.01	-0.15	0.05	-0.07	0.0	-0.19	0.12	0.21	-0.04	0.16	0.14	-0.03
Journalists are concerned about criticism they get from their audiences.	3.35	-0.38	0.02	-0.08	-0.20	0.13	0.28	-0.07	-0.72	0.13	0.65	0.05	-0.27	0.17	0.27

Tab. 1: Attitudes towards Journalistic Responsibility and Audience Criticism [Question: "Please evaluate the following statements. Please rate on a scale from 1 (I totally disagree) to 5 (I fully agree)."]

In this case, the survey results display both transcultural commonalities and contradictions. One of the commonalities is the insight that traditional non-state means of media regulation are apparently much less influential than state-related instruments. The data show that most of the discussed self-regulatory instruments are ascribed a moderate or weak impact only: Press and media councils reach an average of 2.96—and are thus only little more effective than media journalism in the daily news (2.73) or ombudspersons (2.32). These findings confirm our initial assumption that the old system of media self-regulation is indeed facing a crisis—and that alternative approaches are desperately needed. At least, professional codes of ethics (3.44) and company editorial guidelines (3.74) are distinctly more influential. The latter is in fact the only instrument of media accountability that reaches a higher rating than the item "Laws regulating the media" (3.70) which has been included for the sake of comparison. Taken together, the results make it presumable to believe that traditional instruments of media self-regulation—across all journalistic cultures that are covered by this study—are only perceived as effective if—in the case of a breach of journalistic rules—they are associated with noticeable consequences for the liable actor. From the perspective of the respondents, this preconditions seems to be realized primarily in the instances of normative instruments such as legal or editorial rules, but less in the case of press councils or ombudspersons, which lack the power of sanctions for the most part.

However, our survey also reveals notable differences between the analyzed countries, which seem to result—at least partially—from the varying degrees of journalistic professionalization. Thus, for instance, Finland (+1.13) and Switzerland (+0.24) have been having press and media councils since decades, which are now well anchored within the profession and therefore received ratings clearly above the average. On the other hand, a lack of institutionalized media self-regulation is particularly evident in countries like Poland (-0.69) or Romania (-0.68), where journalists concede only little potential to this type of institution—similar to their colleagues in Austria (-0.39), who had to witness how their national press council was incapacitated for as long as a decade, after a conflict with the *Kronen-Zeitung* in 2002.

A country-specific comparison of the acceptance of professional and organizational codes also displays interesting deviations: Our survey data show quite clearly that journalists from Northern and Western European countries (e.g. Finland: +0.92; United Kingdom: +0.52; Switzerland: +0.39) ascribe a markedly higher impact to professional ethics codes, whereas their colleagues in the Mediterranean area (Italy: +0.70; France: +0.33) as well as in Central and Eastern Europe (Estonia: +0.45; Romania: +0.45; Poland: +0.42)—where several journalists' associations with contrasting codes of ethics are competing for supremacy—are rather emphasizing the efficacy of organizational and editorial guidelines. In these latter countries, the newsroom constitutes—at least potentially—the central space for the implementation of journalistic rules. Exemptions in this context are represented by the interrogated journalists for example from Germany and Spain, whose answer patterns distinctly differ from the results of neighboring countries in the same journalism culture. These deviations demonstrate that the cultural differences with regard to media self-regulation and accountability indeed seem to correlate at least in part with the system types by Hallin and Mancini, but in some respects they also diverge.

In comparison to these traditional instruments of media (self-)regulation, what significance do participatory online instruments of media regulation have? A first glimpse at our survey data (Tab. 2) shows that these instruments do not exert an overwhelming impact either. Web-based user comments reach an overall mean of 2.84, thus at least outflanking the influence of media journalism after all. However, media criticism on the various platforms of the social web (2.61) and citizen blogs as a means for media observation (2.25) already fall short of most of the other instruments. For this reason, it becomes clear very quickly that the undisputed potentials of online communication are in no way the new miracle cure in the general struggle for more journalistic responsibility.

However, illuminating differences are detectable here as well: For example, it is striking that citizen blogs and media criticism on the social web are by trend judged to be more influential in those countries without a long tradition of media self-regulation. This applies most notably to our Arab countries, Jordan and Tunisia, where these two items jut out. One of the reasons for the high acceptance of participatory instruments in these countries might be that journalists perceive the audience as the most important entity of objective criticism, while institutions like "media councils" have often been (or still are) an instrument which the regimes used to monitor disliked journalists. In the Arab countries, therefore, participatory media regulation could indeed become a motor for media accountability processes that would not come to unfold without the inclusion of the user perspective. In most of the other analyzed countries, however, the attitude towards participatory media regulation hardly deviates from the overall average.

	Means	Finland	Netherlands	Germany	Austria	Switzerland	United Kingdom	France	Italy	Spain	Estonia	Poland	Romania	Jordan	Tunisia
Company editorial guidelines	3.74	-0.54	-0.05	-0.39	-0.56	-0.12	-0.06	0.33	0.70	-0.88	0.45	0.42	0.45	0.23	0.13
Laws regulating the media	3.70	0.58	-1.01	0.23	0.06	0.22	0.66	0.06	-0.65	-0.25	0.04	-0.04	-0.53	-0.02	-0.12
Professional codes of ethics	3.44	0.92	-0.09	-0.21	-0.21	0.39	0.52	0.20	-0.66	-0.32	0.56	-0.52	-0.61	-0.44	0.23
Press council	2.96	1.13	-0.27	0.10	-0.39	0.24	-0.06	-0.18	0.75	-0.15	0.18	-0.69	-0.68	-0.08	-0.18
User comments	2.84	0.67	-0.45	0.27	-0.02	0.14	0.12	-0.21	-0.48	0.35	-0.08	-0.06	-0.15	-0.12	-0.26
Media journalism	2.73	0.89	0.25	-0.10	-0.10	-0.21	0.13	-0.25	-0.52	0.08	-0.01	-0.20	-0.05	0.17	0.28
Criticism on social media	2.61	0.09	-0.11	0.13	0.20	0.07	-0.35	-0.23	-0.09	-0.10	0.14	-0.04	0.06	0.50	0.33
Ombudsman/Readers' editor	2.32	-0.08	0.09	-0.04	-0.02	0.12	0.26	-0.06	-0.49	0.48	-0.50	-0.13	-0.23	0.21	-0.02
Citizen blogs	2.25	0.20	0.05	0.05	-0.08	-0.07	-0.15	-0.26	-0.32	0.50	-0.07	-0.08	-0.36	0.40	0.54

Tab. 2: Impact of Selected Media Accountability Instruments [Question: "Which of the following have most impact on journalists' behavior in [country]? Please rate on a scale from 1 (No impact at all) to 5 (Very high impact)."]

As a matter of course, the general skepticism towards most of the traditional instruments of media self-regulation as well as towards different web-based channels of media criticism, which is indicated by our survey data in manifold cases and on various levels, raises new questions: Why is it that neither professional nor participatory media accountability instruments unfold a broad influence on practical newsroom work? An advanced analysis of our dataset suggests that an explanation might be found in the ethical orientation of journalistic actors. Indeed, the questioned journalists almost unanimously subscribe to the statement that responsibility is a prerequisite for press freedom. Questioned to whom they felt responsible as a journalist, however, they mostly state: "My conscience"—even more than "Journalistic standards" (Tab. 3). Whereas the own conscience reaches a total mean of 4.78 on a scale from 1 to 5 (with 1 meaning "I don't feel any responsibility towards" and 5 meaning "I feel high responsibility towards"), journalistic standards received an average rating of at least 4.49. Journalists' sources (4.53) are almost equally important as a point of reference of journalistic responsibility. Democratic values (4.23), target audiences (4.31), and the public in general (4.12) follow closely; the internal hierarchy—colleagues (3.72), editors-in-chief (3.76), CEOs/publishers (3.28)—is already by far less important for journalists. Advertisers (1.92), political parties/ideas (1.49), and the government (1.58), as well as religious authorities/groups (1.62) and God (1.93) do not have much significance in this context. From a transcultural perspective, these results show that journalistic performance is still primarily aligned with individual values, while professional values are a point of reference with only secondary importance.

This general trend basically applies to all of the journalism cultures that are covered by the MediaAcT sample. However, major and minor differences between the analyzed countries are detectable with regard to some of the external influences that have an effect on journalistic responsibility. The impact of religious beliefs on media performance is particularly conspicuous in the survey results from Jordan where the ratings for God (+2.60) and religious authorities/groups (+1.15) as points of reference for journalistic responsibility lie notably above the average. The influence of political entities (parties, government) is also considerably stronger in Jordan, but just as well in Tunisia and some of the countries from Eastern (e.g. Romania) and Southern Europe (e.g. Spain) where the systems of journalism and politics are not (yet) completely detached. Furthermore, economic factors—for example in the form of advertisers—are much more likely to shape journalistic responsibil-

	Means	*Finland*	*Netherlands*	*Germany*	*Austria*	*Switzerland*	*United Kingdom*	*France*	*Italy*	*Spain*	*Estonia*	*Poland*	*Romania*	*Jordan*	*Tunisia*
My conscience	4.78	0.04	0.09	-0.06	0.01	0.02	-0.03	0.04	0.11	0.0	0.02	-0.10	-0.03	-0.08	-0.04
My sources	4.53	-0.02	0.04	-0.01	-0.11	0.02	0.18	0.14	-0.14	-0.03	0.0	-0.09	0.06	0.02	-0.25
Journalistic standards	4.49	0.20	-0.21	0.05	0.03	0.22	0.16	0.20	-0.56	-0.27	-0.09	-0.04	-0.01	0.07	0.0
My target audience	4.31	0.17	-0.33	0.04	-0.25	-0.17	0.06	-0.18	0.22	0.20	-0.01	-0.23	0.17	0.33	-0.02
Democratic values	4.23	-0.24	-0.31	0.28	0.48	0.06	-0.20	0.09	-0.05	0.04	-0.10	-0.23	-0.10	-0.04	-0.03
The public in general	4.12	0.10	-0.36	-0.30	-0.30	-0.07	0.04	0.10	0.49	0.22	-0.01	-0.24	0.18	0.35	-0.15
My editor-in-chief	3.76	0.38	0.25	-0.16	-0.23	0.21	0.26	0.06	-0.08	-0.04	0.19	-0.07	-0.10	-0.16	-0.75
My colleagues in the newsroom	3.72	-0.16	0.15	-0.03	-0.19	0.20	0.47	0.07	-0.33	-0.03	0.32	-0.10	0.11	-0.39	-0.74
My CEO/publisher	3.28	-0.13	-0.38	-0.12	-0.21	-0.05	0.22	-0.08	-0.01	0.36	-0.23	0.36	0.38	0.15	-0.44
God	1.93	-0.48	-0.33	-0.12	-0.42	-0.71	-0.38	-0.54	0.08	-0.29	-0.12	0.34	1.41	2.60	–*
Our advertisers	1.92	-0.05	-0.38	-0.02	-0.11	-0.13	0.11	-0.45	-0.07	0.64	-0.22	0.09	0.16	0.60	-0.01
Religious authorities/groups	1.62	-0.36	-0.13	-0.13	-0.04	-0.39	0.06	-0.11	-0.14	0.17	-0.18	-0.10	0.42	1.15	0.15
The government	1.58	-0.17	-0.05	-0.27	-0.21	-0.17	-0.05	-0.33	-0.38	0.45	0.57	-0.09	0.19	0.91	0.49
Political parties/ideas	1.49	-0.25	-0.04	-0.21	-0.19	-0.08	0.04	-0.19	-0.30	0.43	0.04	-0.17	0.22	0.89	0.33

Tab. 3: Points of Reference of Journalistic Responsibility [Question: "To whom do you feel responsible as a journalist? Please rank on a scale from 1 (I don't feel any responsibility towards) to 5 (I feel high responsibility towards)." (=no data available)]*

ity in the Arab world as well as Eastern and Southern Europe as compared to most of the countries in Western Europe that have a longer history of press freedom and journalistic professionalization. All in all, the impact of these media-external points of reference of journalistic responsibility is, however, still comparably small.

On the whole, it is interesting to note that the audience (be it the specific target audience of a journalistic medium or the public in general) is a slightly more influential factor in countries like Jordan (+0.33 and +0.35, respectively), Italy (+0.22 and +0.49), Spain (+0.20 and +0.22), or Romania (+0.17 and +0.18), according to our survey data. This finding can be interpreted as further support for the assumption that participatory instruments of media regulation might be a particularly valuable innovation in countries without a formalized system of traditional media self-regulation.

Conclusion

The survey by the MediaAcT consortium is the first comparative communicator study to provide systematically collected empirical data on journalists' attitudes towards media self-regulation and accountability. But what do these data tell us about the status quo and the future prospects of different media accountability instruments in the digital age? And what potentials do they ascribe to the concept of journalistic quality management as crowd wisdom that recent technological changes have rendered possible?

First of all, the survey shows that the aim of journalistic responsibility is regarded as a key element of a free and pluralistic media landscape. The interviewed journalists displayed little disagreement with regard to this fundamental insight, no matter which journalism culture they were socialized in. At the same time, however, most of the instruments that are supposed to guarantee responsible journalism seem to have only little to moderate impact on journalistic behavior. This somewhat deplorable result applies both to traditional institutions of media self-regulation—such as press councils or ombudspersons—and to new web-based tools for media watching—such as media blogs or user criticism on the social web. According to the survey data, one reason for the general weakness of media self-regulation and participatory media regulation can be found in the strong individual ethical orientation of journalistic actors that overlays professional journalistic values and makes the journalist's own conscience more important than any other point of reference of media responsibility.

Despite undisputed shortcomings, the idea of fostering web-based processes of media observation as a point of origin for journalistic quality management is not completely erroneous either: The survey has also demonstrated that the concept of participatory media regulation may prove particularly valuable in those countries that cannot look back on a long history of journalistic professionalism and media self-regulation. Here, audience criticism fills the void that originated from a lack of other non-state means of holding the media to account. In this way, the Internet can turn into a motor for media responsibility—if the journalists are ready to listen to what their new cyber watchdogs have to say.

Nonetheless, the most effective instruments of media regulation are those with a direct power to sanction journalistic misbehavior. Next to media laws, the respondents in the survey highlighted the impact of in-house editorial guidelines, above all. Apparently, they are regarded as an influential factor because a breach can have direct consequences for the wrongdoer—for example if he loses his job. Interpreted like this, the survey can be read as an appeal to probe more deeply into the possibilities of the newsroom as a setting for media accountability processes—an undertaking that would need further empirical research to provide reliable data. For now, two practical recommendations seem to be more or less self-evident conclusions from the MediaAcT survey (Fengler et al. 2014):

First of all, as media accountability instruments on the organizational level seem to be a promising option, journalists should start thinking about adequate quality initiatives within their own newsroom. By openly endorsing media accountability and responsibility, they will not only improve the quality of their journalistic output, but also (re)gain the trust of their audiences—a currency that is ever more important in a time when many media outlets lament a loss of subscribers which causes serious financial problems as well. The MediaAcT project has collected a large variety of best-practice examples for such initiatives in the form of a *Best-Practice Guidebook* that is available online.[2] Journalistic practitioners are sure to find lots of stimulating ideas in this booklet which can help them to redefine their idea of quality management in the digital age.

2 http://www.mediaact.eu/fileadmin/user_upload/Guidebook/Best_Practice_Guidebook_new.pdf, 02/20/2014.

Secondly, special attention needs to be paid to the transition countries, for example in Eastern Europe or in the Arab world, where democracy and press freedom are relatively new achievements—or even not yet a reality. As many of these countries do not have functional institutions of media self-regulation so far, the integration of participatory processes of media regulation into practical newsroom work opens up many chances. The further stimulation and development of a culture of media responsibility and accountability is also a task for (international) media policy-makers. For the development of sustainable policy strategies, the research by the international MediaAcT consortium offers manifold thought-provoking impulses.

Works Cited

Baum, Achim et al. (eds.) (2005). *Handbuch Medienselbstkontrolle*. Wiesbaden: VS Verlag für Sozialwissenschaften.

Bertrand, Claude-Jean (2000). *Media Ethics & Accountability Systems*. New Brunswick, London: Transaction Publishers.

Beuthner, Michael, and Stephan Alexander Weichert (eds.) (2005). *Die Selbstbeobachtungsfalle. Grenzen und Grenzgänge des Medienjournalismus*. Wiesbaden: VS Verlag für Sozialwissenschaften.

Buschow, Christopher et al. (2013). Social TV in Deutschland—Rettet soziale Interaktion das lineare Fernsehen? *MedienWirtschaft*, 10 (1), 24–32.

Buttler, Günter, and Norman Fickel (2002). *Statistik mit Stichproben*. Reinbek bei Hamburg: Rowohlt.

Campbell, Angela J. (1999). Self-regulation and the Media. *Federal Communications Law Journal*, 51, 711–772.

Eberwein, Tobias (2010). Von "Holzhausen" nach "Blogville"—und zurück. Medienbeobachtung in Tagespresse und Weblogs. In Tobias Eberwein and Daniel Müller (eds.). *Journalismus und Öffentlichkeit*, 143–165. Wiesbaden: VS Verlag für Sozialwissenschaften.

Eberwein, Tobias et al. (eds.) (2011). *Mapping Media Accountability—in Europe and Beyond*. Köln: Herbert von Halem Verlag.

Eberwein, Tobias, Tanja Leppik-Bork, and Julia Lönnendonker (2013). Participatory Media Regulation: International Perspectives on the Structural Deficits of Media Self-regulation and the Potentials of Web-based Accountability Processes. In Manuel Puppis, Matthias Künzler, and Otfried Jarren (eds.). *Media Structures and Media Performance*, 135–158. Wien: Verlag der Österreichischen Akademie der Wissenschaften.

Eberwein, Tobias et al. (2014). Counting Media Accountability—The Concept and Methodology of the MediaAcT Survey. In Susanne Fengler et al. (eds.). *Journalists and Media Accountability. An International Study of News People in the Digital Age*, 65–79. New York: Peter Lang.

Elia, Cristina (2007). Vierzig Jahre Presseombudsmann: Wer sind die Leserschaftsanwälte und wie kommunizieren sie? Eine vergleichende Analyse. *Zeitschrift für Kommunikationsökologie und Medienethik*, 1, 100–105.

Engesser, Sven (2008). Partizipativer Journalismus: Eine Begriffsanalyse. In Ansgar Zerfaß, Martin Welker, and Jan Schmidt (eds.). *Kommunikation, Partizipation und Wirkungen im Social Web. Band 2: Strategien und Anwendungen: Perspektiven für Wirtschaft, Politik und Publizistik*, 47–71. Köln: Herbert von Halem Verlag.

Evers, Huub, Harmen Groenhart, and Jan van Groesen (eds.) (2010). *The Newsombudsman: Watchdog or Decoy?* Diemen: AMB Publishing.

Fengler, Susanne (2003). Holding the News Media Accountable. A Study of Media Reporters and Media Critics in the United States. *Journalism & Mass Communication Quarterly*, 80 (4), 818–832.

Fengler, Susanne (2008). Media Journalism and the Power of Blogging Citizens. In Torbjörn von Krogh (ed.). *Media Accountability Today … and Tomorrow: Updating the Concept in Theory and Practice*, 61–67. Göteborg: Nordicom.

Fengler, Susanne (2012). From Media Self-regulation to "Crowd-criticism": Media Accountability in the Digital Age. *Central European Journal of Communication*, 2, 175–189.

Fengler, Susanne et al. (eds.) (2014). *Journalists and Media Accountability. An International Study of News People in the Digital Age*. New York: Peter Lang.

Fengler, Susanne et al. (2014). Summary: Perspectives for Newsrooms, Policy-makers and Journalism Educators. In Susanne Fengler et al. (eds.). *Journalists and Media Accountability. An International Study of News People in the Digital Age*, 265–288. New York: Peter Lang.

Hallin, Daniel C., and Paolo Mancini (2004). *Comparing Media Systems. Three Models of Media and Politics*. Cambridge: Cambridge University Press.

Heikkilä, Heikki et al. (2012). *Media Accountability Goes Online. A Transnational Study on Emerging Practices and Innovations*. MediaAcT Working Paper, 14, 02/20/2014, http://www.mediaact.eu/fileadmin/user_upload/WP4_Outcomes/WP4_Report.pdf.

Hutter, Andres (2009). *Watchblogs: Medienkritik 2.0? Eine inhaltsanalytische Untersuchung journalistischer Qualität in medienkritischen Weblogs*. Boizenburg: Verlag Werner Hülsbusch.

Jempson, Mike (2013). A Right Royal Tussle between Press Freedom and State-sanctioned Self-regulation. Will a New UK Model Have Anything to Offer Europe? In *MediaAcT. Final Research Report*, 5–7. Dortmund: Erich Brost Institute for International Journalism.

Loosen, Wiebke, and Jan-Hinrik Schmidt (2012). (Re-)discovering the Audience. The Relationship between Journalism and Audience in Networked Digital Media. *Information, Communication & Society*, 15 (6), 867–887.
Mayer, Florian L. et al. (2008). Watchblogs aus der Sicht der Nutzer: Befunde einer Onlinebefragung zur Nutzung und Bewertung von Bildblog. *Media Perspektiven*, 11, 589–594.
McQuail, Denis (2013). *Journalism and Society*. London: Sage.
Meier, Klaus, and Julius Reimer (2011). Transparenz im Journalismus. Instrumente, Konfliktpotentiale, Wirkung. *Publizistik*, 56 (2), 133–155.
Porlezza, Colin (2006). Bashing the Competition, Indulging in Self-adulation. In Sylvia Egli von Matt, Cristina Elia, and Stephan Russ-Mohl (eds.). *Media Journalism in the Attention Cycle. Problems, Perspectives, Visions*, 47–52. Lugano: Giampiero Casagrande.
Pöttker, Horst (2003). Zahnloser Tiger? Plädoyer für wirksame Selbstkontrolle des Journalismus im Dienste der Kommunikationsfreiheit. In Langenbucher, Wolfgang R. (ed.). *Die Kommunikationsfreiheit der Gesellschaft. Die demokratischen Funktionen eines Grundrechts*, 379–384. Wiesbaden: Westdeutscher Verlag.
Pöttker, Horst (2010). Balance of Powers. Medienräte zwischen Selbst- und Fremdkontrolle. *Communicatio Socialis*, 43, 282–298.
Przeworski, Adam, and Henry J. Teune (1970). *The Logic of Comparative Inquiry*. New York: Wiley.
Psychogiopoulou, Evangelia (ed.) (2012). *Understanding Media Policies. A European Perspective*. Basingstoke: Palgrave Macmillan.
Puppis, Manuel (2009). *Organisationen der Medienselbstregulierung. Europäische Presseräte im Vergleich*. Köln: Herbert von Halem Verlag.
Schönhagen, Philomen (1995). *Die Mitarbeit der Leser. Ein erfolgreiches Zeitungskonzept des 19. Jahrhunderts*. München: Reinhard Fischer.
Schönherr, Katja (2008). Medienwatchblogs als Form journalistischer Qualitätskontrolle. In Ansgar Zerfaß, Martin Welker, and Jan Schmidt (eds.). *Kommunikation, Partizipation und Wirkungen im Social Web. Band 2: Strategien und Anwendungen: Perspektiven für Wirtschaft, Politik und Publizistik*, 116–133. Köln: Herbert von Halem Verlag.
Springer, Nina (2011). Suche Meinung, biete Dialog? Warum Leser die Kommentarfunktion auf Nachrichtenportalen nutzen. In Jens Wolling, Andreas Will, and Christina Schumann (eds.). *Medieninnovationen. Wie Medienentwicklungen die Kommunikation in der Gesellschaft verändern*, 247–264. Konstanz: UVK.
Stapf, Ingrid (2006): *Medien-Selbstkontrolle. Ethik und Institutionalisierung*. Konstanz: UVK.
Starck, Kenneth (2010). The News Ombudsman: Viable or Vanishing? In Tobias Eberwein and Daniel Müller (eds.). *Journalismus und Öffentlichkeit*, 109–118. Wiesbaden: VS Verlag für Sozialwissenschaften.

Surowiecki, James (2005). *The Wisdom of Crowds. Why the Many are Smarter than the Few and How Collective Wisdom Shapes Business, Economies, Societies and Nations.* New York: Anchor.

Theis-Berglmair, Anna Maria (2009). Nachrichtenselektion und Leserfeedback. Ein kontingenzorientierter Ansatz zur Analyse von redaktionellen Entscheidungen. *Medien & Kommunikationswissenschaft*, 57 (3), 316–335.

Wied, Kristina, and Jan Schmidt (2008). Weblogs und Qualitätssicherung. Zu Potenzialen weblogbasierter Kritik im Journalismus. In Thorsten Quandt and Wolfgang Schweiger (eds.). *Journalismus online—Partizipation oder Profession?*, 173–192. Wiesbaden: VS Verlag für Sozialwissenschaften.

Wohn, D. Yvette, and Eun-Kyung Na (2011). Tweeting about TV. Sharing Television Viewing Experiences via Social Media Message Streams. *First Monday*, 16 (3), 02/20/2014, http://firstmonday.org/ojs/index.php/fm/article/view/3368/2779.

Ziegele, Marc et al. (2013). Männlich, rüstig, kommentiert? Einflussfaktoren auf die Aktivität kommentierender Nutzer von Online-Nachrichtenseiten. *Studies in Communication | Media (SC|M)*, 1, 67–114, 02/20/2014, http://www.scm.nomos.de/aktuelles-heft-und-archiv/2013/heft-1/.

Quality in Journalism Reconsidered: The Limits of Realism

Mitchell Stephens[1]

Most Americans today think of journalists the way most journalists think of themselves: as reporters of news. Indeed, many practitioners and patrons of traditional journalism in the United States would accept the standard for the profession enunciated by Bill Keller in 2009: "By quality journalism I mean the kind that involves experienced reporters going places, bearing witness, digging into records, developing sources, checking and double-checking." (Keller 2009)[1] Keller was then the top editor of the *New York Times*.

This article argues that this view of journalism, which arrived in the nineteenth century, is outdated in the twenty-first. It will propose an alternative understanding of quality in journalism. And, in so doing, the text will also suggest an alternative response to journalism's crisis.

Yes, the crisis. In recent decades, things have not been going well for journalists: Although the population has continued to grow, total newspaper circulation in the United States has dropped more than 25 percent since 1990, according to the Pew Research Center. The evening newscasts of the three traditional broadcast television networks in the United States—once the dominant news sources in the country—have lost more than half their audience since 1980. Meanwhile, two of the country's major newsmagazines no longer publish regularly, while the third, *Time*, watched its newsstand sales plummet 27 percent in 2012 alone.[2]

Large numbers of talented, hardworking people have, as a result, lost their jobs. Newspapers in the United States employed about 30 percent fewer journalists in 2012 than they did at the beginning of the twenty-first cen-

---

1 This is an excerpt from Stephens, Mitchell (2014). *Beyond News: The Future of Journalism*. New York: Columbia University Press. Copyright © 2014 Columbia University Press. Published with Permission of Paul & Peter Fritz AG (Zürich), N° 70.063.

2 These numbers are based on data or estimates through 2012 (Pew 2013). I have also consulted the Pew Research Center's Project for Excellence in Journalism (Pew 2011).

tury, according to that Pew report.[3] Online and cable news operations have grabbed audiences but not replaced enough jobs—not, at least, enough that pay. A particularly chilling consequence of these cutbacks has been a drop in the number of professional reporters patrolling city halls, statehouses and the sites of our recent or lingering wars.[4]

In their gloom, journalists and those who value them have been repairing to conferences to rend their clothes, wail their wails and curse the Fates. Some, overcoming a much-prized dispassion, pen apocalyptic screeds and mouth dreadful prophecies.

These defenders of the craft—in their anxiety, in their distress—are wont to declare news itself to be in crisis.[5] In this, however, they are wrong. News, for the most part, is in fine shape.

News can be defined as new information about a subject of some public interest that is shared with some portion of the public.[6] It is hard not to notice that the amount of new information of public interest being shared continues to swell even as jobs handling it vanish. Indeed, the recent arrival of the most powerful information technology in human history has, as one might have expected, been a boon for news. The Web remains very young, but already it gathers accounts of an extraordinarily wide variety of events from an extraordinarily wide variety of sources and disseminates them, in a wide variety of formats, fast and far. Amazingly fast: Newsworthy events nowadays—statements, contretemps, disasters—are all over the Web before they are over.[7] And impressively far: Our computer-Internet-cell-Web devices, despite the efforts of some governments to restrict them, now zip words and images in and out of almost every country on earth. We have not before seen a news medium like this.

---

3 Pew's source here is the American Society of News Editors, Newsroom Employment Census, 2010; but Pew (2013) has calculated and added estimates for more recent years.

4 See Dorroh 2009. "In 2003, there were more than a thousand foreign journalists covering the war in Iraq," noted Jill Abramson, then managing editor of the *New York Times*, in 2010, when foreign troops were still fighting there. "Today that number has dwindled to fewer than one hundred" (Abramson 2010). That 2013 Pew Center report complains of "a continued erosion of news reporting resources."

5 See, for example, this panel on "the Crisis in News"—a panel at the University of California, Berkeley in 2008: http://fora.tv/2008/04/26/The_Crisis_in_News_Print_Media .

6 This is the definition I introduced in *A History of News* in 1988.

7 For an interesting Heideggerian perspective on technological developments that promote the "live" and the "real time," and the distortions they cause for "eventization"; see Stiegler 1998, 16 and elsewhere.

This does not mean the news that comes to us on our laptops, smart phones, tablets or, soon, wearable devices is always edifying, constructive or reliable. News in print or on television, after all, has often enough failed to embody those virtues. And the Web's manifold strengths as a news medium do not mean all news will be equally well served by it. As amateurs and algorithms collect and distribute more news, issues of accuracy, accountability, bias, depth and professionalism loom larger. We have to be alert, as we must be with any medium, for blind spots. Once, it should be remembered, journalism reviews devoted themselves to cataloging the many egregious oversights of newspapers and newscasts—with their sometimes narrow-minded and conventional "gatekeepers."

The Web's weak points, at first glance, appear to fall into two categories: One is a failure regularly to report on news that grew up with and, perhaps, grew old with newspaper beat systems: varieties of local news in particular. The second is a reluctance to deploy teams of experienced reporters, of which the Web so far has few, and devote space, of which it has plenty, to original investigations.

However, the multitudes that have by now obtained Internet connections have access to a bountiful supply of news. The gates have flung open. And the flow of news on the Web and its cousins seems, if anything, likely to continue to broaden, deepen and accelerate. Entrepreneurs and nonprofits have even begun having a go at some of those blind spots. The future of news, in other words, appears reasonably secure.

It is the future of journalism that is looking grim.

Journalism's Current Problems

Defining exactly what journalism is has proven surprisingly difficult (Splichal and Sparks 1994, 1–31). It is, in some sense, a specialized enterprise—usually performed for pay—but define it too narrowly and you risk depriving non-specialists, amateurs, of its privileges and responsibilities (Calvert 1999). It is clearly about news, but again too narrow a focus on reporting and distributing the news—Bill Keller's focus in 2009—can limit the possibilities of this enterprise. Here is my definition: Journalism is the activity of collecting, presenting, interpreting or commenting upon the news for some portion of the public.

For approximately the last century and a half, journalists have emphasized the collecting and presenting of news. Since the middle of the nineteenth century, journalists have made their living—indeed constructed grand enterprises—either through the sale of news or through the sale of ads next to news. It does not look as if similar numbers of journalists will be able to make their living this way in the future.

In an effort to coax some money out of their growing numbers of online-only customers, many news organizations have begun constructing often partially permeable "pay walls"[8]. Still, the flood of information on current events now sloshing around the Internet can be sipped, drunk or bathed in on hundreds of news sites, blogs, feeds, apps—thousands for international stories—mostly for free. This has, understandably, strained profit margins and flummoxed business models—including pay walls—at many of those news sites, blogs, feeds and apps. There is not much of a living in hawking that which is given away free. That is like running a ferry after they've built the bridge. It is like selling ice after everyone has a refrigerator. Ask encyclopedia salesmen today, if you can find one.

Radio and television newscasts had overcome this problem with advertising. However, unlike the supply of programs on radio or TV back in their heyday as advertising media, the supply of news-rich pages on the Internet is now so large that it is hard to charge much for ads on those pages.[9] Even the supply of online audio and video news, which can be preceded by little commercials, seems to be heading in the direction of unlimited, as audio and video become ever easier to record, edit, upload and access. The success of digital news, consequently, itself undercuts the economics of digital journalism: because it is so plentiful it is worth less to audiences and advertisers. Now that information pours out of spigots, the point is, painstakingly gathered facts on current events—what happened, who said what, when—have lost much of their value.

Probably the Associated Press, Reuters, Bloomberg and a few additional news organizations will be able to make a go of supplying written or video accounts of such events to websites fast and in bulk. However, it is difficult to see many others basing a business on the fruits of such run-of-the-mill reporting—on merely collecting and presenting routine news. After more

8 For the spread of pay walls, see the discussion of newspapers in Pew Research Center 2013.
9 For a discussion of advertising's problems online, see Wolff 2012.

than a century and a half of selling the latest facts, journalists need to sell something else.

Technologies and "Quality"

With a desperation characteristic of people whose livelihoods are at stake, journalists have been forced in recent years to rethink how "quality journalism" might be distributed: Newspapers and magazines contemplate abandoning print; radio-news programs produce podcasts; television-news networks have begun to think of themselves as video-report suppliers. And it is difficult to find a news operation today that does not arm its reporters with a variety of recording devices and that is not splashing news up on Facebook, Twitter and some spiffy new app for Apple's latest insanely desirable iDoohickey. Journalists today are even rethinking how "quality journalism" might be funded: perhaps through nonprofits, if not pay walls.

Few today, however, are rethinking what, in these changing times, "quality journalism" might be. Yet definitions of quality, like technologies, are vulnerable to obsolescence. Consider—to jump fields and centuries—the case of Ernest Meissonier.

Meissonier, who died in 1891, was for a time the most respected painter in Paris and, therefore, the world (King 2006). "He is the incontestable master of our epoch," gushed another master painter, Eugène Delacroix. Delacroix proclaimed to Charles Baudelaire that "amongst all of us, surely it is he who is most certain to survive" (King 2006, 2).

Meissonier excelled at precision and accuracy: "I paint like everyone else," he once explained. "Only I look always" (Hungerford 1999, 39). His work was based, in other words, on intensive observation—of the exact positions adopted, for example, by a moving horse: "What efforts, what sketches, what lengths of precious time, what fatigue he incurred," one contemporary exclaims, "to faithfully translate the living animal" (Gotlieb 1996, 165)! Critics raved about Meissonier's "scrupulous exactitude" (Hungerford 1999, 162, 173). His painstakingly precise recreations of great events dominated the most important expositions and commanded the highest prices.

Meissonier's masterwork, which took him twelve years to complete, was a depiction of Napoleon and his army at Friedland in 1807. It was sold to an American for 380-thousand francs in 1876.[10] *Friedland* is currently on display in New York's Metropolitan Museum of Art and features quite a few well-realized horses.

Ernest Meissonier saw himself as part of a "tradition of [...] honesty, conscientiousness and truthfulness" (Hungerford 1999, 162). But by the end of the nineteenth century such efforts to get every detail just right were starting to seem a waste of time, and the tradition Meissonier exemplified was starting to seem something of a bore. A new technology was helping to change understandings of what painting might be.

Photography demonstrated, to begin with, that the "truthfulness" offered by Meissonier's paintings was limited—even when it came to horses' gaits. The American photographer Eadweard Muybridge, employing speeded-up cameras and a system of trip wires, had managed to photograph horses in motion. Muybridge, who was quite aware of Meissonier's work, confirmed the painter's depiction of the walk, but Muybridge's photos showed different leg positions during the trot and, in particular, the gallop than Meissonier had been able to produce by merely looking—however intently.[11] Horses legs simply moved too fast for naked eyes—even Meissonier's eyes—to follow. "If I could only repaint *Friedland*," Meissonier, who was quite aware of Muybridge's work, was reported to have said (Hungerford 1999, 205). Meissonier's later paintings restrict themselves to horses that are walking or standing still (Hungerford 1999, 206).

And this new technology not only beat realistic painters at their own game; it did so, as shutter speeds continued to accelerate, without all the "efforts," "sketches" and "fatigue." Initially the camera, with its revelations, contributed to a fever for facts from which realistic painters like Meissonier benefited. But soon enough photography devalued precisely that at which Meissonier specialized: It made producing painstakingly accurate recreations of just about anything easy and, therefore, cheap.

Attitudes were, in time, adjusted accordingly. In the late nineteenth and early twentieth centuries the notion that quality in art was dependent upon precision and verisimilitude faded. Ernest Meissonier's reputation tumbled

10 Ross King, *The Judgment of Paris*, 369. See also, for the price, Henry James (1989, 108). Constance Cain Hungerford (1999, 174), however, gives the price as 300-thousand francs.

11 Marc J. Gotlieb (1996, 183) explains that "Muybridge's stills forced the collapse at the end of the nineteenth century of a verist standard of artistic truth that had prevailed in Salon painting for a half-century or more."

after his death to the point where one major two-volume history of French art in the nineteenth century did not mention his name. The Louvre eventually exiled a marble statue of Meissonier from its halls (King 2006, 370–371).

Have technologies today, particularly technologies introduced in the past couple of decades, done the same with the painstaking gathering of information on current events? Through blogs, emails and tweets, audiences can themselves receive firsthand reports on newsworthy happenings, as those events trot or gallop along. Through video and live cameras, audiences can often even "observe" those events themselves. Have technologies today, therefore, beaten traditional news reporting at its own game? By making it so easy for audiences to find out what happened, have they reduced the value of telling them what happened? Have they, in other words, outdated the view of quality in journalism many traditional journalists still hold: This veneration of witnessing, digging, finding sources and checking?

Bill Keller, striking something of an apocalyptic tone himself, lamented the "diminishing supply" of his version of "quality journalism" (Keller 2009). But "diminishing supply" can sometimes be a sign of diminishing need. The Web grants us all the ability to dig, find sources and check—to *search*, in a word. It often grants us the ability to witness. Is it possible the supply of the kind of journalism in which this is done for us, like the supply of realistic painting after photography, *should* diminish? Maybe we don't require as many of Keller's "experienced reporters" as we did before much of the world presented itself to us online.

Without making light of the professional crisis journalism has experienced, an opportunity, at least, can be discerned here—for journalism. For the Web allows our best journalists to surrender the prosaic task of telling everyone what just happened. It allows them to leave the speeches and press conferences in part to the cable networks and YouTube; to leave the interviews with police and survivors to diligent wire service reporters; to fob off surveillance of some lesser boards and councils on civic-minded individuals with an urge to tweet and post. It would be a blow to what sometimes seems a colonial conception of reporting, but our best journalists might even surrender some of the responsibility for filling in Americans on what has been going on in India, Egypt or Iraq to cosmopolitan residents of India, Egypt or Iraq. It might no longer be quite so necessary to parachute in a jaded American with a notepad. The Web allows our best journalists—it requires them, I will argue—to return to an older and higher view of their calling: not as reporters of what's going on but as individuals capable of providing a wise take on what's going on.

Certainly—at difficult times, in hard-to-reach places—"bearing witness" can reveal and expose. Certainly, facts must be checked, worried over. Certainly, much still can be uncovered by "digging into records" and "developing sources." Nothing said here is meant to imply that journalists should forgo human-rights reporting, investigative reporting, enterprise reporting or exclusives.

However, merely noting who said what at a some public event—mere "stenographic" reporting, to use the most dismissive of the terms that has been applied to it—shares some of the flaws today of a Meissonier painting: It is occasionally less starkly truthful than simply watching or listening to recordings of that event, and it is less revealing than freer, more far-reaching discussions of that event online by well informed individuals with an insight or point of view to share.

The Last Bastion of Realism

Ernest Meissonier's work was connected, of course, to a movement that extended well beyond painting: A movement proclaimed by Honoré de Balzac in the preface to one of his books in 1830s: "The author firmly believes that details alone will henceforth determine the merit of works" (Balzac 2009). Details dominated much nineteenth-century fiction—realistic, naturalistic, even coarse, often morally or politically troubling, details. Though his work was not coarse or troubling, Meissonier, like Balzac, was a details man. So, for another example, was the English physician John Snow who, in the middle of that century, located cholera deaths and wells on a map to demonstrate that the disease spread through contaminated drinking water.[12] The pursuit of details, of facts, was crucial to the advance of nineteenth-century science

12 The cholera story is well told in Steven Johnson (2006). David Mindich emphasizes the shift from religious explanations for cholera in the American press to more fact-based explanations: "What more fitting way is there to enter the age of 'realism' than with a replacement of God with the 'foul effluvia of [...] overflowing privies.'" The latter is a quote from the New York Tribune; David T. Z. Mindich (1998, 111). Mindich connects the effort to prevent cholera to realism in general and the rise of objectivity in the press in particular in this book.

and medicine. And, as that century progressed, increasing numbers of journalists became details men and women, too.

But, despite the great accomplishments of what Balzac had called the impulse to "venerate the fact" (Stephens 2007, 246), literature, painting and, in some sense, science would move on in the next century. They would cease to imagine the world as composed of discreet, independently verifiable details, which could be captured by sufficient observation, "conscientiousness and truthfulness." Novelists, painters and, in some sense, scientists left realism behind in the twentieth century.

Most journalists in the United States did not. To the contrary: Their embrace of realism—of facts, of details, conscientiously and truthfully rendered—became ever more passionate in the twentieth century.[13]

As early as 1859 the poet Charles Baudelaire was mocking the view that "Art is, and cannot be other than, the exact reproduction of Nature" (Stephens 1998, 75). The Impressionists, later in that century, were much inclined toward nature, but they didn't necessarily look for it where the Meissoniers of the world had found it. "Nature is not on the surface," Paul Cezanne proclaimed. "It is in the depth" (Doran 2004, 124). In the first decades of the twentieth century, paintings would appear that showed aspects of nature, human faces in particular, from multiple perspectives at once or that showed nothing easily recognizable as nature. In 1919 Virginia Woolf, writing about fiction, was attacking the "materialists" and noting how "life escapes" their "magnificent apparatus for catching" it (Woolf 1984, 285–286). Painting and literature turned from diligent attempts to capture material details to impressions and perspectives; they turned from the objective to the subjective. Physics turned, in important ways, from a universe filled with the fixed and knowable to a view of even the physical world, at great enough speeds or small enough sizes, as relative and uncertain.

Journalism missed that turn. In the United States, only a limited number of journalism critics[14] and a small minority of journalists in the twentieth century questioned the merit and scope of the "scrupulous exactitude"

13 David Mindich (1998, 5) writes: "It is no less than remarkable that years after consciousness was complicated by Freud, observation was problematized by Einstein, perspective challenged by Picasso, writing was deconstructed by Derrida and 'objectivity' was abandoned by practically everyone outside newsrooms, 'objectivity' is still the style of journalism that our newspapers articles and broadcast reports are written in, or against."

14 For an example of a journalism critic who did question realism, see Robert Karl Manoff (1986).

reporters were getting better and better at pursuing.[15] In 1922 even Walter Lippmann, among the wisest of journalists and journalism critics, thought "reality" could be easily located, although, not surprisingly, he had some difficulty settling on exactly where it might be found—"outside" us, not in the subjective "pictures in our heads" or, as he writes some pages later, in "the interior scene" not the "façade" we present to the world (Lippmann 1922; Stephens 1991).

By the start of the twenty-first century, journalists had grown a little more humble. "Journalism does not pursue truth in an absolute or philosophical sense," acknowledged one distinguished group, the Committee of Concerned Journalists, itself administered by the Project for Excellence in Journalism. But in its "Principles of Journalism" this group did insist that journalism "can—and must—pursue" truth "in a practical sense." It asserted, in a somewhat clumsy wording, that "this 'journalistic truth' is a process that begins with the professional discipline of assembling and verifying facts." And the Committee of Concerned Journalists was not at all humble in its view of the importance of such verified and assembled facts: "Democracy depends on citizens having reliable, accurate facts put in a meaningful context" (Pew undated).[16]

Some of the words used here—"practical," "process," "context"—bespeak an awareness that piling detail upon detail does not a world make. That is encouraging. But in their pursuit of "excellence" such committees of journalists still have some difficulty surrendering the nineteenth- and twentieth-century notion that journalists are primarily collectors of facts.

15 The "New Journalists" of the last third of the twentieth century waged more of a revolution in writing style than a revolt against the primacy of detail. Indeed, Tom Wolfe (1973, 34) perhaps the most eager of this bunch to debate with journalists, fashioned himself a great partisan of realism: "The introduction of realism into literature by people like Richardson, Fielding and Smollett was like the introduction of electricity into machine technology." Nonetheless, among the freedoms introduced by the "New Journalism" was the freedom to interpret.

16 'Principles of Journalism.' Pew Research Center's Project for Excellence in Journalism, 03/07/2014, http://www.journalism.org/resources/principles. "These became," Pew explains, "the basis for *The Elements of Journalism*, the book by PEJ Director Tom Rosenstiel and CCJ Chairman and PEJ Senior Counselor Bill Kovach"—an influential book by two influential thinkers on journalism.

Journalism and Modernism

Which is odd, since newspapers, newsreels and newscasts, though allergic to modernism and postmodernism, rank high among the forces that spurred modernism and postmodernism in the twentieth century. They helped outmode simpler conceptions of reality through their relentless interrogations, their compulsive cynicism, their mixings of high and low, their unavoidable heterogeneity and their incessant reminders of the importance of the way something is communicated.

"Is the press a messenger?" asks the Viennese critic Karl Kraus in 1914. "No it is the event itself." Kraus' characteristic bitterness, after the start of the First World War, exceeds even his characteristic irony. The press, he complains, is "the most murderous weapon" progress wields, for it destroys the nineteenth-century assumption that an occurrence might have an existence independent of the way we talk about it—the assumption, even, that the occurrence proceeds the way we talk about it: "Deeds are stronger than words," Kraus writes, "but the echo is stronger than the deed. We live on the echo, and in this topsy-turvy world the echo arouses the call." (Kraus 1998, 204–205)

The press spent much of the twentieth century demonstrating the power of "the echo"; yet many journalists refused to take the lesson. Stiff-necked and unblinking in their own focus on events, lost in their own chase after deeds, spellbound by the perpetrators of such deeds, many journalists failed to note the debt those events and deeds owed to culture, to language, to dissemination. They continued to believe that the facts they were "assembling and verifying" had an existence entirely independent of the process of assembling, verifying, investigating, sorting and presenting in "a meaningful context."

Walter Cronkite always closed the *CBS Evening News,* probably the leading news source in the United States in the late 1960s and 1970s, by announcing, "And that's the way it is." He did not say—and I have seen no evidence that he thought—"and that's one perspective on the way it is." And Cronkite certainly did not say—and I would be surprised if he or most of his contemporaries in journalism thought—"and that's the way our coverage helps make it seem to be."[17]

---

17 I am not arguing that Walter Cronkite (1998) intended his tagline, "And that's the way it is," as some sort of philosophical statement. He stumbled upon it, by his own account. And this statement was criticized at the time by CBS News President Richard Salant

In the nineteenth century, the best journalists had seen themselves as, and were often welcomed as, participants in literature and philosophy.[18] In the twentieth century they seemed, many of them, at war with literature and philosophy. Indeed, journalists would sometimes be dismissed as philistines. For they had become champions of precisely the era and ideology contemporary artists and thinkers were aggressively positioning themselves against: Journalism in the United States in particular, outside of the arts pages at least, became the last redoubt of realism—of the insistence that there was, if you could get the facts right, a "way it is." "Objectivity" was American journalism's religion. Its practitioners transformed themselves into such fierce defenders of realism that they sometimes came through with parodies of avant-garde art or experimental prose, and dismissals of literary theory as "mumbo-jumbo" (Stephens 1991). Indeed Tom Wolfe—before he switched to novels among the most adventurous and creative of journalists—became one of the leading critics of the turn away from realism in literature and art (Wolfe 1975; ibid. 2009).

Many of the ways and means of realism remained beneficial, of course, for journalism: Going places is good, digging and checking are good; so are thoroughness, industry and fairness. Let us not underestimate the accomplishments of the realism that lived on in journalism, in a century desecrated by some terrible and mighty events and deeds. The list of exposés managed by reporters in the United States in the twentieth century—with their doggedness in pursuit of old-fashioned, nineteenth-century-style facts—is long and impressive: It includes Lincoln Steffens on municipal corruption, John Hersey on the horrific consequences of the atomic bomb dropped on Hiroshima, Rachel Carson on the dangers of pesticides, Seymour Hersh on an

for its presumption, a criticism Cronkite understood. But the presumption in question, according to Cronkite, was the assumption that the broadcast would prove to have been "mistake" free—not that its perspective on reality was necessarily subjective or limited, let alone that it was making news through its reporting of the news. Also see Jeff Alan (2003, 129–130) and Karen S. Johnson-Cartee (2005, 114).

18 Patrick Collier (2006, 201) writes: "English newspapers and literature lived happily together for 200 years. [...] So closely tied were literature and journalism that the distinction between them would have made no sense to writers from Daniel DeFoe [...] to Matthew Arnold. [...] By the age of modernism, such a productive nexus of literature, journalism and politics seemed unimaginable." Collier is writing of England, but this close tie also existed in America and its severing, which Collier credits more to quality than ideology, may have been even more dramatic here. The relationship between Henry James and the New York *Tribune* is representative of that tie, its break and my understanding of what went wrong.

American atrocity in Vietnam, and Bob Woodward and Carl Bernstein on foul doings in the Nixon White House. Tom Wolfe also had a point in arguing that, with many novelists having departed for more subjective realms, journalists had something of an exclusive on some contentious and colorful decades in the last half of the twentieth century (Wolfe 1975, 28–35). The "magnificent apparatus" for capturing facts constructed by traditional journalism in the United States has often performed impressively.

And it has, often enough, seemed sufficient. It was hard to argue that journalists needed a Pablo Picasso or a Virginia Woolf, let alone a Werner Heisenberg, to tell us about the latest presidential campaign. But wouldn't that coverage have benefited, as it attempted to disentangle that which had been spun, from a little more facility with perspective, a somewhat more extensive acquaintance with the wanderings of the human mind and a surer grasp of the limits of certainty? As the founder and editor of the *Talking Points Memo* website, Josh Marshall, noted during the 2012 U.S. presidential primaries, "Campaigns have a way of not working by logical principles or, perhaps we should say, Newtonian physics" (Marshall 2012). And didn't Woolf do rather well with the First World War? Didn't Picasso do rather well with Guernica? For details alone—even if "reliable," "accurate" and placed in "a meaningful context"—have serious limitation even as guides to the world of current events. The case can be made that politics sometimes "escapes" this journalism, as does much of the rest of "life."

Interpretation

In his journal even Delacroix, though one of Ernest Meissonier's biggest admirers, confided his suspicion that "there is something else in painting besides exactitude and precise rendering" (King 2006, 226). There is something else in journalism, too.

Jill Abramson succeeded Bill Keller as the top editor of the *New York Times* in 2011. In 2010 she had offered her own explanation of what "quality journalism" provides: "trustworthy information about the world we live in—information that is tested, investigated, sorted, checked again, analyzed and presented in a cogent form" (Abramson 2010, 39–44). This was, with one exception, compatible with Keller's understanding of "quality journalism." But that exception is significant. For Abramson includes a word not used by her predecessor: "analyzed." That is a word with which a more up-to-date conception of journalism might begin.[19]

The definition of journalism introduced in this book is careful to acknowledge that journalism might be more than "exactitude and precise rendering." That definition includes interpreting or commenting upon the news—as bloggers today and printers at the time of the American Revolution would expect it to include; as our editorial writers, columnists and contributors to opinion journals and others serious magazines would expect it to include. For even at the height of American journalism's fidelity to realism, even in the middle of the twentieth century, some examples of first-rate interpretation and comment could be found—on certain pages, in certain publications.

The Committee of Concerned Journalists, Bill Keller and most traditional journalists are, no doubt, aware that there are other journalisms besides "the discipline of assembling and verifying facts." I do not believe they meant to exclude interpretation and comment entirely from the realm of quality journalism. Keller's *New York Times* did after all devote two pages each day specifically to opinion. (Before and after he was the paper's executive editor, Keller's own essays appeared on one of those pages.) And the prominence of "news analysis" pieces on other pages in the paper increased on Keller's watch. After she replaced Keller, Abramson seems to have begun going further: deemphasizing somewhat the *Times'* efforts to present its readers with all the important details on yesterday's most important events.[20]

In fact, interpretation and comment are popping up all over journalism today. Yes, commentary can often be shrill and predictable—on talk radio, on cable television, on many websites. But some of this interpretation and comment is original and enlightening—particularly in our better magazines, particularly on our smarter blogs and websites, sometimes in the *New York Times*. The loud and the shrill get the attention of journalism critics. But

19 In this article, Abramson (2010, 39–44) seems confident of the importance of "quality journalism"—a term she uses eight times. But she sometimes seems to be struggling to settle upon the proper relationship between information and analysis in "quality journalism": "People don't crave just information. They seek judgment from someone they can trust, who can ferret out information, dig behind it and make sense of it. They want analytic depth, skepticism, context and a presentation that honors their intelligence."

20 Because it is the best and most influential of our traditional news organs, any criticism of current thinking on journalism must take on the *New York Times*. Its efforts, in print and online, are often be discussed in these pages. The *Times* has been changing with the times. My argument is that, despite its many strengths, it needs to change more.

original and enlightening forms of analytic journalism are pointing the way to the future.

This chapter is not calling for the invention of something new. But it is calling for us to aspire, with more awareness and understanding, to a journalism that *regularly* does much more than simply recount who said or did what yesterday. It is calling for a broader, more ambitious journalism. And it is arguing that such a journalism will not fully arrive without a change in mindset.

Wisdom Journalism

Journalism has undergone enormous changes at other times. The inventions that spur such periods of change are hard to miss, as are the methodological improvements, particularly greater speed, those exciting new machines bring. The upheavals such technological advances impose upon the business of journalism are also seen and grasped without much delay: old enterprises begin to stumble and fall; new enterprises, led by unorthodox innovators, soon call attention to themselves. However, the realization that *understandings* of journalism must also change spreads more slowly—particularly for those who had achieved success with the old understandings, particularly for the orthodox. New mindsets are needed to best employ the new technologies and save some of the old businesses. But mindsets inevitably lag behind.

Consider for another example of a mainstream twenty-first-century journalist's mind that of Martin Baron, who took over as the top editor of the *Washington Post* in January 2013. Months earlier Baron had delivered an inspiring speech to newspaper editors in which he had advocated openness to "innovation" and all sorts of change—innovation and change *except* in the newspaper's "traditional purpose and […] core values."

And what is that "traditional purpose"? Baron's answer, a good summary of a kind of journalistic orthodoxy, began: "Stories that are well reported. Stories that are well told. Stories that offer a window on the communities we serve." "Insights" were mentioned but only the sort that arrive, according to the familiar journalistic trope, "from looking beneath the surface"—from digging, from reporting, in other words. "Interpretation" or "analysis" did not appear among the one hundred and twenty seven words Baron used to describe what he believes ought to be preserved in journalism.

Baron emphasizes, in this speech, the importance of "doing all of this work—all of this journalism—honestly, honorably, accurately, and fairly." He does not mention, among these "core values," the importance of doing journalism knowledgeably, intelligently or insightfully (Romenesko 2012). Indeed, the notion that journalists today might have to offer understandings not just facts was not considered at any point in Baron's more than three-thousand-word speech.

Marty Baron, like the other journalists I have been quoting, is smart, accomplished and public spirited. Neither he nor they are unaware of the value of analytic journalism. In 2005, while Baron was editor of the *Boston Globe*, a series in that newspaper exploring the science and ethics of stem-cell research won the Pulitzer Prize for "Explanatory Journalism." But when Baron, upon leaving the *Globe*, was asked what he was proud of at that newspaper, he mentioned, in a *Globe* reporter's paraphrase, "its investigative reporting, arts coverage, narrative journalism, and war reporting in Iraq and Afghanistan, as well as its push into digital and multimedia formats." Baron apparently neglected, once again, to mention interpretation or analysis (Weisman 2012). Mainstream journalists often neglect to mention them.

Indeed, the number of journalists today who continue to overlook the importance of interpretation is evidence that the rethinking of journalism—the transformation of the traditional journalist's mindset—has just begun. It is, therefore, not surprising that the remaking of journalism has a long way to go, too.

My argument is that quality in twenty-first century journalism lies not so much in "experienced reporters going places, [...] developing sources, checking and double-checking"—though all those activities are certainly useful. My argument is that quality in journalism lies instead in what I am calling "wisdom journalism"—journalism that strengthens our understanding of the world.

Wisdom journalism is an amalgam. It includes, to begin with, the more rarified forms of reporting—exclusive, enterprising, investigative. Though they remain in short supply, there is no controversy about the worth of such original forms of reporting. Much of what Marty Baron is proud of at the *Boston Globe* would qualify. But wisdom journalism also includes, even emphasizes, informed, interpretive, explanatory, even opinionated takes on current events.[21]

Much more of this kind of writing, audio, video or Web work is needed. It must continue to expand well beyond a few serious magazines, a bunch of thoughtful blogs and those two pages, a Sunday section and some "news analysis" pieces in newspapers like the *New York Times*. It needs to become mainstream. We need to develop standards to help distinguish such wisdom journalism from the shrill and predictable. And we need to get better at it.

That will require not only a change in mindset but further and substantial changes in the way journalism is currently practiced: If providing insight rather than gathering facts is to become a main mission—not an auxiliary mission—traditional methods of hiring and promoting might no longer apply. The assignment structure will have to be transformed.

Evenhandedness and dispassion might no longer be dominant values. Original perspectives will push widely available information from whatever passes for a front page. And we will need to apply as much thought to standards for journalistic argumentation as we have applied to how to cover an event.

This sounds like a new approach to journalism. Actually it is in some ways an old one—from the days before American journalism's enthrallment with realism.

Works Cited

Abramson, Jill (2010). Sustaining Quality Journalism. *Daedalus*, 139 (2), 39–44.

Alan, Jeff (2003). *Anchoring America*. Chicago: Bonus Books.

Balzac, Honoré de (2009). *The Chouans*. New York: Collier.

Calvert, Clay (1999). And You Call Yourself a Journalist? Wrestling With a Definition of 'Journalist' in the Law. *Dickinson Law Review*, 103 (Winter 1999), 411–451.

Carey, James W. (1986). The Dark Continent of American Journalism. In Robert Karl Manoff and Michael Schudson (eds.). *Reading the News*, 146–196, New York: Pantheon.

21 In his thoughtful discussion of the civic consequences of the twenty-first-century crisis in journalism, Alex S. Jones (2011, 3) distinguishes between the "news of assertion" and "fact-based news" or the "news of verification," and Jones notes that "traditional journalists have long believed that this form of fact-based accountability news is the essential food supply of democracy." I am arguing, in part, for the "news of assertion."

Collier, Patrick (2006). *Modernism on Fleet Street.* Hampshire: Ashgate Publishing Limited.

Cronkite, Walter (1998). On His 'That's the Way It Is' Signoff. Interview. *Archive of American Television*, 03/07/2014, http://www.youtube.com/watch?v=NOa4sg2WOEQ.

Doran, Michael (2002). *Conversations with Cézanne.* Berkeley: University of California Press.

Dorroh, Jennifer (2009). Statehouse Exodus. *American Journalism Review*, 03/07/2014, http://www.ajr.org/article.asp?id=4721.

Gotlieb, Marc J. (1996). *The Plight of Emulation.* Princeton: Princeton University Press.

Hungerford, Constance Cain (1999). *Ernest Meissonier: Master in His Genre.* Cambridge: Cambridge University Press.

James, Henry (1989). *The American Purchase of Meissonier's* Friedland. The Painter's Eye. Madison: University of Wisconsin Press.

Jones, Alex S. (2011). *Losing the News: The Future of the News that Feeds Democracy.* New York: Oxford University Press.

Johnson, Steven (2006). *The Ghost Map: The Story of London's Most Terrifying Epidemic—and How It Changed Science, Cities and the Modern World.* New York: Riverhead Books.

Johnson-Cartee, Karen S. (2005). *News Narratives and News Framing.* Oxford: Rowman & Littlefield.

Keller, Bill (2009). Talk to the Newsroom: Executive Editor. *New York Times*, January 30, 03/07/2014, http://www.nytimes.com/2009/01/30/business/media/02askthetimes.html?pagewanted=all&_r=0.

King, Ross (2006). *The Judgment of Paris.* New York: Walker & Company.

Kraus, Karl (1999). From *In These Great Times.* In Vassiliki Kolocotroni et al. (eds.). *Modernism: An Anthology of Sources and Documents.* Chicago: University of Chicago Press.

Lippmann, Walter (1922). *Public Opinion.* New York: Harcourt, Brace and Company.

Manoff, Robert Karl (1986). Writing the News (by Telling the 'Story'). In Robert Karl Manoff and Michael Schudson (eds.). *Reading the News*, 227–228, New York: Pantheon.

Marshall, Josh (2012). Sounds Right. *Talking Points Memo*, 03/07/2014, http://talkingpointsmemo.com/archives/2012/01/sounds_right.php?ref=fpblg.

Meyer, Philip (2008). The Elite Newspaper of the Future. *American Journalism Review*, 03/07/2014, http://www.ajr.org/article.asp?id=4605.

Mindich, David T. Z. (1998). *Just the Facts.* New York: New York University Press.

Pew Research Center (undated). *'Principles of Journalism.' Pew Research Center's Project for Excellence in Journalism*, 03/07/2014, http://www.journalism.org/resources/principles.

Pew Research Center (2011). The State of the News Media: An Annual Report on American Journalism, 03/07/2014, http://stateofthemedia.org/2011/.

Pew Research Center (2013). The State of the News Media, 03/07/2014, http://stateofthemedia.org/2013/overview-5/.

Romenesko, Jim (2012). *Boston Globe Editor: Newspapers Are Badly Bruised, But Not Beaten*, 03/07/2014, http://jimromenesko.com/2012/09/17/boston-globe-editor-newspapers-are-badly-bruised-but-we-are-not-beaten/#more-24850.

Splichal, Slavko, and Colin Sparks (1994). *Journalists for the 21st Century*. Norwood, N.J.: Ablex.

Stephens, Mitchell (1991). Deconstruction and the Get-Real Press. *Columbia Journalism Review*, September/October 1991, 38–42.

Stephens, Mitchell (1998). *the rise of the image the fall of the word*. Oxford University Press: New York.

Stephens, Mitchell (2007). *A History of News*. 3rd ed. New York: Oxford University Press.

Stiegler, Bernard (1998). *Technics and Time, Vol. 1*. Stanford: Stanford University Press.

Weisman, Robert (2012). Martin Baron, Editor of *The Boston Globe*, to Become Editor of *The Washington Post*. *Boston Globe*, November 13, 03/07/2014, http://www.boston.com/businessupdates/2012/11/13/martin-baron-editor-the-boston-globe-become-editor-the-washington-post/ycJWyd1XOsuvpQAQ3LMStI/story.html.

Wolfe, Tom (1973). The New Journalism. In Tom Wolfe and Edward Warren Johnson (eds.). *The New Journalism*, New York: Harper Row.

Wolfe, Tom (1975). *The Painted Word*. New York: Picador.

Wolfe, Tom (2009). *From Bauhaus to Our* House. New York: Macmillan.

Wolff, Michael (2012). The Facebook Fallacy. *Technology Review*, May 22, 03/07/2014, http://www.technologyreview.com/web/40437/?nlid=nldly&nld=2012–05–22.

Woolf, Virginia (1984). Modern Fiction. In Andrew McNeille (ed.). *The Common Reader*, 146–154, New York: Harcourt.

User Participation and Professional Journalism on the Internet: Theoretical Background and Empirical Evidence

Christoph Neuberger

Introduction

The "digital revolution" has many facets. Perhaps the most important change, which happened in the last two decades was the lowering of the entrance threshold to the public sphere (Neuberger 2009). On the Internet almost everyone can be a provider or a communicator. The so-called "mass media" press and broadcasting has only opened the door for the masses on the audience's side. The door has remained closed for most people on the speakers' side of mass communication. Technical barriers (e.g. frequency shortage) and economic barriers (high costs for production and distribution) restrict the number of broadcast stations and press companies. The publishing capacity of the traditional mass media was low—calculated on the basis of the number of press offerings and channels or the number of printed pages and the volume of broadcast time. In the past professional journalists were the powerful "gatekeepers" who controlled the access to the public sphere. A widespread criticism in regard to this is that professional journalism privileges elite members of society or gives priority to its own interests. The mass audience had few possibilities for feedback in the era of "gatekeepers."

This bottleneck has disappeared on the Internet. Doors are wide open now—at least in a technical sense. The rise of the "Web 2.0" has made participatory formats of publishing ("social media" like blogs, social network sites and wikis) available, which are easy for amateurs to use. One important consequence here is that journalism is losing its "gatekeeper" role because everyone can talk to everyone else directly. That does not mean that journalism is unnecessary on the Internet, because information must still be filtered and checked for the audience. Journalism is needed as a "gate-watcher" (Bruns 2005) and as a moderator of public debates on the Internet. Yet, must *professional* journalists do this job—why not bloggers or news search engines (Beiler 2013)?

Relationships between Media and Formats

What happened in the mid 1990s was not a clear break in the development of the media and journalism. Old media and professional journalism do still exist. And it does not seem, as if their further existence were only a transient state—briefly before they disappear. It may also be the case that the Internet is another step in an ongoing differentiation of public communication where many media and formats do co-exist and do not substitute each other. What we find at the moment in the field of journalism is the "contemporarity of the incontemporary." This observation necessarily raises the general question: What is the relationship between old and new (1) media and (2) formats of public communication?

(1) Media: The question whether old media (like the newspaper) and new media (like the Internet or earlier "new" media like television) compete with or complement each other was often discussed and explored in empirical studies (as a research overview: Mögerle 2009). A hundred years ago Wolfgang Riepl (1913[1972]) formulated the law of complementarity (also called *'Rieplsches Gesetz'*), which says that an old medium will never be substituted by a new medium if it finds a niche, where it does not compete with a superior medium. The third relationship between media is cross- media. That means that outlets in different media (press, broadcasting, Internet) carry the same brand name (e.g. *BBC, The New York Times, Der Spiegel*) and are often produced by the same newsroom staff. The outlets support each other, if they are able to guide the audience from one outlet to another.

(2) Formats: Media are not the subject of this paper. It focuses on the relationships between the formats of professional journalism and the formats of participatory communication on the Internet. Here almost the same relationships can be distinguished: competition, complementarity and integration (Neuberger and Nuernbergk 2010, 320; Neuberger et al. 2013).

- *Competition:* Do professional journalism and social media compete with each other? Do they offer equal gratifications for users? And do they achieve the same level of quality as professional journalism does from the user's point of view? There has been an intense debate on whether blogs are doing better or worse than journalism. Don Alphonso (2004, 26), a well-known German blogger, once stated that blogs would be the coffin nail of the traditional journalism. Critics like Andrew Keen (2007) stress the incompetence of amateurs, which have no professional training. Social media must be functionally equivalent to legacy media to gain the status of a competitor. Results of empirical studies show typical strengths (like authenticity, subjectivity and interactivity) and weaknesses (like the absence of newsworthiness, objectivity, credibility, transparency and neutrality) of the so-called "participatory journalism" or "citizen journalism" on the Internet (see as a research overview Engesser 2013).

- *Complementarity:* Complementarity means that professional journalism and social media are different entities but affect each other in positive or negative ways. Complementary relationships between them are for example the gaining of traffic and user attention for media websites from social media, and vice versa. Journalists investigate stories and conduct research, using social media. On the other hand, blogs and other social media usually adopt topics from journalism and mirror the media agenda to a large extent. Journalists also monitor audience reactions and follow up communication to their reports on the Internet. Another complementary relationship is the criticism of journalism in "watch blogs."
- *Integration:* A third relationship is the integration of participatory formats into media websites. In that case they become part of professional journalism. *Twitter*, *Facebook*, *Google+*, *YouTube* and blogs are multifunctional tools for journalism: Journalists promote their own websites when they refer and link to the content of their site. Journalists also write columns in blogs or conduct real-time coverage on *Twitter*, directly from where the events are taking place. Finally, journalists use social media to interact with the members of their audience in various ways (Boczkowski and Mitchelstein 2012). In the role of the "producer" they even support the production of news content (Bruns 2008), for example as a "reader reporter" (*'Leserreporter'*) or "iReporter" (Kperogi 2011) who sends photographs or videos to the editorial office.

To sum it up, we expect complex relationships between professional journalism and the users of social media. Most studies are restricted to a single medium or format. Therefore, we are actually just at the beginning of a deeper understanding of how different media and formats are interwoven and how they interact.

Earlier Formats of User Participation in Journalism

Before the emergence of the Internet one-way and one-to-many communication were the dominant modes in the public sphere. Yet, the call for participation in journalism, and its realization have a long history. Letters to the editor have been the most prominent way of interaction between journalists and their readers for centuries (Schönhagen 1995; Heupel 2007; Wahl-Jorgensen 2007; Mlitz 2008; Sehl 2013). Jürgen Habermas (1990, 105) described a network of many local publics in which (male, educated and wealthy) citizens debated face-to-face in coffee houses and other places about literature and politics. These small circles were connected through early magazines (*'Moralische Wochenschriften'*). The links between the different levels of the public sphere (Habermas 1992, 452) allowed a mutual exchange of ideas. The formation of opinion was not only a top-down, but also a bottom-up motion. Habermas stated that the mass media of the 19th and 20th century dissolved this close connection and two-sided flow of information between journalism and its audience. This is not entirely correct because letters to the editor and reader debates have maintained their importance especially in many local newspapers (*'Heimatzeitungen'*) (Schönhagen 1995; Mlitz 2008, 148–204). The Internet reunites the many small publics and the large public. That is the idea of the "networked public sphere" (Benkler 2006). The Internet integrates public communication in the vertical dimension when it adds a "long tail" (Anderson 2007) to the mass communication, where especially users of social media can be found, who only gain limited reach and resonance. The question is whether the flow of information and the formation of opinion are actually more open, diverse and balanced in the new virtual world than in the old world of mass media. Critics, such as Hindman (2009), say that on the Internet a hierarchy with a few high influentials, who receive a lot of attention and a majority of almost unnoticed writers, has been established. Here it can be argued that journalism should be a promoter of user participation that gives ordinary citizens the chance to be heard ("public journalism") (Haas 2007). And journalism should also be the moderator of public discourse that ensures that normative expectations will be fulfilled (Edwards 2002).

Relationships between Social Media and Professional Journalism: Evidence from Empirical Research

The Internet is still in flux. The multi-optionality of the Internet and its openness for everyone causes a steady dynamic and enormous heterogeneity of the medium. The Internet has "no shared institutional identity" (McQuail 2005, 139). Because of that, it is very hard in empirical research to generate results which can be generalized for the Internet as a whole, and which are more than a snapshot. Studies with a comparative (e.g. Singer et al. 2011) and historical perspective are still missing. Our picture of the Internet landscape is still very fragmented.

In the following sections results from several empirical studies, which were mostly conducted in Germany in the last years, will be presented. For *competition* to predominate, the audience would have to regard social media as providing essentially similar services. The question of whether this is the case was explored in a user survey. The survey's results will be presented in detail in the fifth section, yet to state a key finding upfront, it is unlikely that journalism is in competition with social media. Because of that, in the sixth section *complementary* aspects will be discussed and the questions will be answered, as to which extent journalistic websites obtain visitors from social network sites, and how journalists make use of social media for research purposes. The seventh section shows how journalists make use of social media and *integrate* them in media websites. In the final section the reasons for the current crisis of professional journalism will be discussed.

Competition: Quality Attributes and Gratifications of Media Websites and Social Media

Which specific features and gratifications do Internet users associate with different formats and services? This was the research question in a study for which 1,000 German Internet users were surveyed between January and March 2011 (Neuberger 2014).[1] The survey participants were recruited in an online panel and were matched with the general distribution of the po-

1 The study was supported by *Deutscher Fachjournalisten-Verband (DFJV)*, Berlin. The field phase was conducted by the marketing and media research institute result GmbH, Köln.

pulation, with respect to age, gender, and educational level. The survey respondents were asked to evaluate formats and services they used at least once every six months, and to checkmark up to three services from a list that embodied a particular feature to the highest degree.

(1) Quality attributes: The results were unequivocal: Among the eleven formats and services that were presented as choices, it was the services provided by professional journalists from the *traditional mass media* and the Internet encyclopedia *Wikipedia,* that matched the individual quality criteria best (Tab. 1). In the opinion of the largest portion of those surveyed, press websites embodied most of the features best. Although *Wikipedia*, television and radio station websites did receive fewer top votes, they were still rated significantly higher than all other services. With the exception of *Wikipedia*, social media was scarcely associated with journalistic features. Only in relation to questions about the author's personal viewpoint and about in-depth discussions did *social network sites* (and weblogs) receive endorsement by the largest portion of those surveyed. Neither of these features is a central characteristic of journalism, however.

In case of *Wikipedia*, it is worth noting that the service does indeed meet the criteria for journalistic quality, but it was primarily valued for its usefulness in daily life and the encyclopedic information it provides. However, *Wikipedia* has none of the identifying features of journalism, such as regular reporting, timeliness, and a broad overview of the news. The fact that *Wikipedia* did better than press websites, with respect to independence, as well as transparency of sources, reveals that the collaborative contributions of lay persons can outperform professional press output in terms of core journalistic quality criteria, at least in the opinion of the users. News search engines like *Google News* performed better than social media and were valued primarily for the broad overview of the news they provide and their timeliness.

(2) User's gratifications: Besides these features, the respondents were also asked about their reasons for using the sites. In a list of 13 gratification items, specific details about information behavior on the Internet were considered. The list differentiated between the passive reception of a pre-produced menu of subjects (surveillance), active search (guidance), accidental reception (serendipity), as well as the user's (meta-) orientation, regarding current information services on the Internet itself. Besides current news information, which is the central focus of journalism, the list also included the acquisition

	Video portals (n=753)	*Twitter (n=197)*	*Social network sites (n=678)*	*Weblogs (n=359)*	*Bookmark collections (n=209)*	*Participatory news platforms (n=307)*	*News search engines (n=762)*	*Web services of newspapers and magazines (n=619)*	*Web services of television and radio stations (n=612)*	*Portals with news (n=662)*	*Wikipedia (n=757)*
Regular reporting	7.0	15.4	13.9	10.0	7.5	10.5	38.8	63.1	57.5	46.4	10.0
Credibility	7.4	9.9	10.5	7.7	6.6	15.0	30.6	57.4	47.8	37.5	43.0
Broad overview of the news	7.6	7.4	9.9	8.0	6.1	17.2	45.9	56.9	48.7	49.1	17.0
Timeliness	12.4	22.0	26.9	9.3	4.2	14.6	45.0	56.6	50.7	49.3	17.1
Objectivity	4.8	5.0	9.0	8.2	6.6	16.2	31.2	56.5	41.4	34.8	51.2
Issue expertise	8.4	8.9	8.6	11.7	4.7	14.2	29.2	53.0	42.8	32.4	40.3
Issues that are important for everyone	11.2	8.0	14.4	7.9	5.6	18.0	37.3	51.5	45.1	40.0	40.3
Usefulness in everyday life	15.0.	11.5	20.3	11.3	7.9	13.6	37.2	47.0	36.3	32.3	54.5
Background information	6.9	12.3	11.1	13.1	8.0	16.8	23.6	44.9	35.8	27.0	41.4

Table 1 continued on next page

	Video portals (n=753)	*Twitter (n=197)*	*Social network sites (n=678)*	*Weblogs (n=359)*	*Bookmark collections (n=209)*	*Participatory news platforms (n=307)*	*News search engines (n=762)*	*Web services of newspapers and magazines (n=619)*	*Web services of television and radio stations (n=612)*	*Portals with news (n=662)*	*Wikipedia (n=757)*
Reader-friendly	17.2	20.4	30.9	20.8	10.3	22.0	25.7	43.8	31.1	31.4	31.6
The author is known	11.8	13.3	18.3	14.2	4.7	12.0	14.2	37.8	24.1	15.0	19.4
Transparency of sources	9.2	7.9	9.6	7.4	7.5	18.8	25.5	36.2	21.0	23.4	46.3
Own research	15.5	12.0	22.3	19.8	6.6	20.4	17.9	34.8	24.1	15.4	34.9
Independence	20.6	17.9	23.2	16.9	10.4	19.7	21.8	26.8	21.1	19.4	34.6
In-depth discussion	19.1	29.8	46.1	32.1	7.1	25.1	12.7	20.3	14.2	18.6	11.0
The author's personal point of view	31.6	27.9	42.1	36.0	8.0	25.1	10.7	16.8	13.5	13.6	22.5

Tab. 1: Characteristics of Internet Formats and Services.
Choice of the formats and services that embody the enlisted quality criteria to the greatest degree; up to three services could be selected. Percentage of those surveyed, who use a service or format at least once every six months (without the response: "I can't say"). User survey, January through March 2011 (Source: Neuberger 2014, 45)

	Video portals (n=753)	*Twitter (n=197)*	*Social network sites (n=678)*	*Weblogs (n=359)*	*Bookmark collections (n=209)*	*Participatory news platforms (n=307)*	*News search engines (n=762)*	*Web services from newspapers and magazines (n=619)*	*Web services from television and radio (n=612)*	*Portals with news (n=662)*	*Wikipedia (n=757)*
To get an overview of current events in *Germany*	4.2	8.9	10.3	7.4	7.0	10.1	34.7	59.6	44.4	40.4	7.2
To get an overview of current events in *my region*	3.4	5.9	15.7	8.0	5.6	9.5	27.1	54.0	28.2	20.8	4.5
To get an overview of current events abroad	4.4	6.9	8.2	8.0	3.8	12.1	34.1	48.2	36.5	36.4	6.3
To find background information about current news topics	5.1	6.0	7.4	7.7	8.0	14.9	32.3	43.4	27.8	25.4	34.8
To be able to converse about important issues	9.4	13.4	26.5	13.5	6.1	13.7	19.2	38.6	27.5	22.6	20.9
To find out about specific subjects of interest to me	9.9	9.5	10.6	13.3	7.5	9.8	40.9	32.1	19.5	20.2	57.0
To stumble upon interesting stories	21.2	15.8	22.2	14.3	8.5	14.0	32.6	29.6	23.5	33.1	19.8

Table 2 continued on next page

	Video portals (n=753)	*Twitter (n=197)*	*Social network sites (n=678)*	*Weblogs (n=359)*	*Bookmark collections (n=209)*	*Participatory news platforms (n=307)*	*News search engines (n=762)*	*Web services from newspapers and magazines (n=619)*	*Web services from television and radio (n=612)*	*Portals with news (n=662)*	*Wikipedia (n=757)*
To learn which Internet sites have information about current issues	5.5	6.9	11.2	12.1	12.1	7.6	44.9	25.4	16.9	25.6	11.5
To search for specific information	7.6	8.5	8.9	9.1	9.3	9.8	50.6	23.8	16.4	20.3	58.2
To entertain myself	47.1	18.0	52.0	13.8	9.4	5.7	12.7	18.1	26.5	16.5	8.3
To have discussions with other people	9.6	19.9	46.4	22.5	4.3	12.5	8.1	13.1	9.8	10.7	5.7
To cultivate my social relationships	8.6	25.8	67.4	9.6	4.2	9.5	6.7	8.4	5.0	10.7	4.6

Tab. 2: Reasons for Using Internet Formats and Services.
Choice of formats and services that are used most often for this purpose; up to three entries are possible. Percentage of respondents who use a service or format at least once every six months (without the response: "No opinion"). User survey, January through March 2011.
(Source: Neuberger 2014, 47)

of basic background knowledge, advice, entertainment, engagement in discussions, and the cultivation of relationships—and thus included areas at the outskirts and edges of journalism.

In relation to the user attribution of features to media formats, the results concerning users' reasons for using the sites turned out quite differently (Tab. 2):

- Users also prefer to engage with legacy media on the Internet if they want to obtain an *overview* about the region, Germany, or foreign news. According to users, professional journalism also defines the topics with which one should be familiar as a citizen. This corresponds to the traditional *gatekeeper* and *agenda-setting* role, ascribed to professional journalism.
- However, if users want to be personally active and are *searching* for specific information or reports about special areas of interest, then they prefer news search engines and *Wikipedia*.
- News search engines and portals with news are especially well suited for *orientating users on the Internet* and the *serendipitous discovery* of interesting subjects.
- The strength of social media lies in *communicative gratification*: Social network sites and weblogs are appropriate sites for discussion, whereas for cultivating relationships, users prefer social network sites and *Twitter*.
- People seeking *entertainment* prefer social network sites and video portals.
- Social media is largely unimportant as a news source (i.e. for an overview of current events). However, social network sites do at least permit people to find out which subjects are important for the sake of conversation. Presumably, this capacity relies upon the fact that it is social network sites that spread the word on subjects that are important in the media.

In summary, the results of the user survey show that people do not lose their consciousness for journalistic quality, when using the Internet. Internet users see sharp differences between formats and services, and in most categories they prefer professional journalism. So it is unlikely that there is an existing competition between journalism, social media, and news search engines.

Complementarity: The Guiding of Users and Journalistic Research

(1) Guiding of users: The dissemination of news story-links on social network sites and other social media outlets has grown increasingly important for journalistic websites, as shown by several recent studies. Purcell et al. (2010) conducted a representative telephone survey in December 2009 and January 2010. The survey found out that 17 percent of Internet users in the U.S. had posted links or thoughts about news stories on social network sites, such as *Facebook* (Purcell et al. 2010, 61). An additional 3 percent posted or re-tweeted a link to an online news story or a blog entry on *Twitter*. Such content recommendations provide a form of orientation, and if users comment on news stories, they become a type of pre-filter for other users. In this way, they help to draw attention to a story, as well as enhance or diminish confidence in the quality of journalistic websites. What impact do such link recommendations have on web traffic flows? How much do they contribute to overall traffic at a particular website?

Olmstead et al. (2011) used New View data from Nielsen to examine how users get to the top 25 news websites in the U.S. They based their analysis on data from the first three quarters in 2010. Nielsen records traffic from a referral site, when at least five users have been linked to it. Data is available for 21 of the 25 top sites.

- On average, 30 percent of the traffic to the top sites came through *Google* (Olmstead et al. 2011, 7–8), referring here to all *Google* services. For 17 of the 21 sites, *Google* was the largest supplier of traffic, and it was the second largest for the four other services.
- Of course, the majority of *Google News* users did not limit themselves to reading the headlines but also clicked on news stories. However, only a few large news websites benefited from this, especially *The New York Times* (15 percent of all users of the *NYT* on the Internet), *CNN* (14 percent) and *ABC News* (14 percent). These three services accounted for 69 percent of all *Google News* users who followed a link to another site (Olmstead et al. 2011, 22). *Google News* promotes the concentration of attention on the Internet.
- *Facebook* was the second or third most important access site for five of the services. Sites that benefited the most from *Facebook* included the

Huffington Post (8 percent), *CNN* (7 percent) and *The New York Times* (6 percent) (Olmstead et al. 2011, 10, 24).
- Surprisingly, the *Drudge Report* was even more significant than *Facebook*: Matt Drudge's online magazine, which became well-known during the Lewinsky scandal in 1998, was the second or third most important user generator in 12 cases (Olmstead et al. 2011, 11).
- However, *Twitter* proved to be quite insignificant (Olmstead et al. 2011, 12).

How often do users go from a news website to a sharing site and share what they have read (Olmstead et al. 2011, 19, 24)? Here, it was mostly *Facebook* that dominated; up to 7 percent of the users (Yahoo News) shared news items on *Facebook* (6 percent for *CNN*; 4 percent for *ABC News*, *Fox News* and *The Huffington Post*; 3 percent for *The New York Times* and *The Washington Post*).

According to an analysis conducted by Comscore in Germany (Schmidt 2011), *Google* Sites and *Google News* lost market share, due to traffic being directed to media websites. By contrast, *Facebook* witnessed a traffic increase. In February 2011, *bild.de* obtained 10 percent of its visits through *Facebook* (February 2010: 2 percent). An increase in visits could also be shown for *spiegel.de* (2011: 4 percent), *focus.de* (3 percent) and *faz.net* (3 percent). Even though the percentage of users who visit a professional journalism website via *Facebook* was in most cases still in the single-digit realm, it appears that recommendations from one's circle of friends and acquaintances are gaining importance as a pre-filter in Germany. (More recent data are not available.)

The possibilities for disseminating and reading news stories over social network sites were documented in a more differentiated way in a non-representative online survey conducted by Wladarsch (2010, 141; Wladarsch 2014). This survey showed that 49 percent of social network site users had already registered comments about media contributions or clicked the *like* button. Of the users surveyed, 34 percent had forwarded media articles themselves, or transferred them from a media website to a social network site.

The presented data shows that the forwarding of news stories through social network sites drives traffic to journalistic websites. Yet, news search engines also make a significant contribution directing users to journalistic websites.

(2) Journalistic research: Another complementary relationship is the journalistic use of social media for research purposes. This is shown, for example, in the findings of a newsroom survey, which collected data from editors-in-chief of Internet news departments in May and June 2010. Survey participants were members of German news departments from superregional media organizations (weekly newspapers and general-interest magazines, broadcasting stations, Internet-only providers) (Neuberger et al. 2011; Neuberger et al. 2013). In the case of daily newspapers, all titles (*'publizistische Einheiten'*) were analyzed, including regional publications. Altogether 70 newsroom directors took part in the survey, reflecting an overall return rate of 45 percent[2]. How often do staff members use computer-based research methods (Tab. 3)? While search engines and web catalogues predominate, social network sites, together with other social media, still represent a relatively small significance.

How often do staff members use the following computer-based research methods?	*Frequently*	*Rarely*	*Never*
Search engines and web catalogues	98.3	1.7	0
Online-services provided by journalistic media	98.3	1.7	0
E-mail	95.0	5.0	0
Internal house archives	83.3	15.0	1.7
Databases and archives	80.0	18.3	1.7
Weblogs	30.0	56.7	13.3
Social network sites	21.7	73.3	5.0
'Social news" services	13.3	75.0	11.7
Twitter	11.7	86.7	1.7
User platforms	8.3	73.3	18.3

Tab. 3: Use of Computer-based Research Methods by Newsroom Staff.
In percent, n=60. Three-step scale (without the responses: "Other" and "I cannot comment on this matter"). Newsroom survey, May through June 2010.
(Source: Neuberger et al. 2011, 60)

2 The study was sponsored by the *Media Authority of North-Rhine Westphalia* (Landesanstalt für Medien Nordrhein-Westfalen, LfM), Düsseldorf.

Integration: Publishing and User Interaction

In the newsroom survey already mentioned, almost every service was used by newsroom staff, in one way or the other (Tab. 4). The question included all sorts of complementary relationships between media websites and social media and their integration into media websites. How important are social media, especially for publishing activities and/or interaction with users (table 5)? The most important services here were *Facebook* and blogs. The focus of the study lay on the use of *Twitter*, made by newsroom staff. Results show that almost all media organizations used *Twitter* to attract readership (97 percent, n=61), and for monitoring audience responses (91 percent, n=58). About two thirds of participants said they used *Twitter* to interact with users (66 percent, n=58) and for live coverage and breaking news (63 percent, n=60) (Neuberger et al. 2011, 46, 55). These data show that all social media have already gained a great importance for journalism. This is not the place to discuss, which specific suitability profile social media have when used by journalists. The following questions remain largely open: What effects has the integration of social media on the quality of journalism? And what can be done to improve its quality?

	Daily newspapers (n=47)	*Weekly newspapers, magazines (n=3)*	*Broad-casting (n=8)*	*Internet-only (n=2)*	*Total (n=60)*
Twitter	95.7	100	100	100	96.7
Weblogs	95.7	100	87,5	100	95.0
Social Bookmarking Services	80.9	100	100	100	85.0
Facebook	85.1	100	100	100	83.3
YouTube	76.6	100	100	100	81.7
StudiVZ/MeinVZ	63.8	100	100	50.0	70.0
MySpace	61.7	100	87.5	100	68.3

Tab. 4: Use of Social Media by Newsroom Staff.
In percent.
Newsroom survey, May through June 2010.
(Source: Neuberger et al. 2011, 64)

	Very important (4)	*Rather important (3)*	*Rather unimportant (2)*	*Unimportant (1)*	*Average*
Facebook (n=53)	45.3	45.3	7.5	1.9	3.3
Weblogs (n=57)	15.8	31.6	38.6	14.0	2.5
Twitter (n=58)	15.5	55.2	25.9	3.4	2.8
Social Bookmarking Services (n=51)	9.8	33.3	41.2	15.7	2.4
StudiVZ/MeinVZ (n=42)	7.1	26.2	45.2	21.4	2.2
YouTube (n=49)	6.1	40.8	40.8	12.2	2.4
MySpace (n=41)	0	2.4	51.2	46.3	1.6

Tab. 5: Importance of Social Media for Newsroom Staff Members, with Regard to Publishing Activities and/or Interaction with Users. Four-step scale. Percentage of respondents who use a service or format (without the response: "Other"). Newsroom survey, May through June 2010. (Source: Neuberger et al. 2011, 66)

Conclusion

The current developments in journalism and in the public sphere are difficult to describe and to evaluate. And it is even more difficult to forecast the future prospects in this field. Though, what we know at the moment is that user participation and professional journalism are not poles apart. They are aligned in many ways. The empirical results presented thus far, could lead to the conclusion that the situation of journalism is in good order, as there is no major competition between social media and media websites. Complementary relationships, from which both sides benefit, dominate. And journalists apply to social media for increased user participation. This harmonious interplay between journalism and its audiences, has been portrayed by Bowman and Willis (2003, 12), in the form of a new media ecosystem model. Yet, as we all know, this image of an *ideal world* is misleading. The Internet has shaken the very economic foundations of professional journalism. And the main reasons for this are not increased user participation and the rise of social media. It is clear however, that professional media providers must bear a substantial share of the responsibility for this: The profusion of choices,

	Average	*Agree completely/ Agree somewhat, in %*
Because I can also obtain adequate information on other sites for free, I am not prepared to pay for journalism on the Internet. (n=938)	2.3	61.4
I come across many news stories on the Internet by chance, so I don't have to search for them specifically. (n=942)	2.7	42.4
In order to be aware of the important topics of the day, I don't have to visit any media websites, since there is also information about them presented and discussed on other sites. (n=918)	2.8	36.3
I find it difficult to evaluate the quality of information on the Internet. (n=956)	3.1	27.6
It's hard for me to distinguish what is journalism on the Internet and what is not. (n=950)	3.1	25.3
When I visit Google News, I only read the headlines and don't call up the whole article. (n=838)	3.1	23.4
Recommended links that I obtain from friends in social network sites are more valuable than search machine results. (n=862)	3.3	20.2
I obtain many news items related to politics and business directly from political party and company websites. Therefore, I don't have to rely on journalism. (n=908)	3.9	13.0
On the Internet, I have the sense that I can't find what I am looking for. (n=950)	3.7	12.2

Tab. 6: Information Behavior and Evaluation on the Internet (1 = agree completely, 5 = disagree completely, without the response "I can't say"). User survey, January through March 2011. (Source: Neuberger 2013, 127)

their omnipresence, their availability free of charge, and the interchangeability of services has diminished brand loyalty and reduced the audiences' willingness to pay for services. This is confirmed by the results of the user survey (Tab. 6). 61 percent of the users are not prepared to pay for journalism on the Internet. Many users do not consider that it is necessary to visit media websites at all, because they find current news in other places, like *Google News*. And a quarter of the users even has difficulties to find out what "journalism" is and what it is not. If journalism is, from a societal perspective, still necessary for news production, navigation and moderation in the public sphere, and if there is no substitute in sight for professional journalism, then journalism must convince its potential customers through better quality and exclusive offerings.

Works Cited

Alphonso, Don (2004). Ein Dutzend Gründe, warum Blogs den Journalismus im Internet aufmischen werden. In Don Alphonso and Kai Phal (eds.). *Blogs! Text und Form im Internet,* 23–43. Berlin: Schwarzkopf & Schwarzkopf.

Anderson, Chris (2007). *The Long Tail. How Endless Choice is Creating Unlimited Demand.* London, Sydney, Glenfield: Random House.

Beiler, Markus (2013). *Nachrichtensuchen im Internet. Inhaltsanalyse zur journalistischen Qualität von Nachrichtensuchmaschinen.* Konstanz: UVK.

Benkler, Yochai (2006). *The Wealth of Networks. How Social Production Transforms Markets and Freedom.* New Haven, London: Yale University Press.

Boczkowski, Pablo J., and Eugenia Mitchelstein (2012). How Users Take Advantage of Different Forms of Interactivity on Online News Sites: Clicking, E-Mailing, and Commenting. *Human Communication Research*, 38, 1–22.

Bowman, Shayne, and Chris Willis (2003). *We Media. How Audiences Are Shaping the Future of News and Information.* Reston, Va.: The Media Center at The American Press Institute, 02/01/2014, http://www.hypergene.net/wemedia/download/we_media.pdf.

Bruns, Axel (2005). *Gatewatching. Collaborative Online News Production.* New York, Washington, D.C., Baltimore: Peter Lang.

Bruns, Axel (2008). *Blogs, Wikipedia, Second Life, and Beyond. From Production to Produsage.* New York, Washington, D.C., Baltimore: Peter Lang.

Edwards, Arthur R. (2002). The Moderator as an Emerging Democratic Intermediary: The Role of the Moderator in Internet Discussions About Public Issues. *Information Policy,* 7, 3–20.

Engesser, Sven (2013). *Die Qualität des Partizipativen Journalismus im Web. Bausteine für ein integratives theoretisches Konzept und eine explanative empirische Analyse.* Wiesbaden: Springer VS.

Haas, Tanni (2007). *The Pursuit of Public Journalism: Theory, Practice, and Criticism.* New York: Routledge.

Habermas, Jürgen (1990). *Strukturwandel der Öffentlichkeit. Untersuchungen zu einer Kategorie der bürgerlichen Gesellschaft. Mit einem Vorwort zur Neuauflage 1990.* Frankfurt am Main: Suhrkamp.

Habermas, Jürgen (1992). *Faktizität und Geltung. Beiträge zur Diskurstheorie des Rechts und des demokratischen Rechtsstaats.* Frankfurt am Main: Suhrkamp.

Heupel, Julia (2007). *Der Leserbrief in der deutschen Presse.* München: Reinhard Fischer.

Hindman, Matthew (2009). *The Myth of Digital Democracy.* Princeton, N.J.: Princeton University Press.

Keen, Andrew (2007). *The Cult of the Amateur. How Today's Internet is Killing Our Culture and Assaulting Our Economy.* London, Boston: Nicholas Brealey Publishing.

Kperogi, Farooq A. (2011). Cooperation with the Corporation? CNN and the Hegemonic Cooptation of Citizen Journalism through iReport.com. *New Media & Society,* 13, 314–329.

McQuail, Denis (2005). *McQuail's Mass Communication Theory.* London, Thousand Oaks, New Delhi: Sage.

Mlitz, Andrea (2008). *Dialogorientierter Journalismus. Leserbriefe in der deutschen Tagespresse.* Konstanz: UVK.

Mögerle, Ursina (2009). *Substitution oder Komplementarität? Die Nutzung von Online- und Print-Zeitungen im Wandel.* Konstanz: UVK.

Neuberger, Christoph (2009). Internet, Journalismus und Öffentlichkeit. Analyse des Medienumbruchs. In Christoph Neuberger, Christian Nuernbergk, and Melanie Rischke (eds.). *Journalismus im Internet: Profession—Partizipation—Technisierung,* 19–105. Wiesbaden: VS.

Neuberger, Christoph (2013). Competition or complementarity? Journalism, Social Network Sites, and News Search Engines. In Heinz-Werner Nienstedt, Stephan Russ-Mohl, and Bartosz Wilczek (eds.). *Journalism and Media Convergence,* 119–129. Berlin, Boston: De Gruyter.

Neuberger, Christoph (2014). Die Identität und Qualität des Journalismus im Internet aus der Sicht des Publikums. Ergebnisse einer Online-Befragung. In Wiebke Loosen and Marco Dohle (eds.). *Journalismus und (sein) Publikum. Schnittstellen zwischen Journalismusforschung und Rezeptions- und Wirkungsforschung,* 229–252. Wiesbaden: Springer VS.

Neuberger, Christoph, and Christian Nuernbergk (2010). Competition, Complementarity or Integration? The Relationship between Professional and Participatory Media. *Journalism Practice,* 4, 319–332.

Neuberger, Christoph, Hanna Jo vom Hofe, and Christian Nuernbergk (2011). *Twitter und Journalismus. Der Einfluss des "Social Web" auf die Nachrichten (LfM-Dokumentation, 38).* Düsseldorf: Landesanstalt für Medien Nordrhein-Westfalen (LfM).

Neuberger, Christoph, Hanna Jo vom Hofe, and Christian Nuernbergk (2013). The Use of Twitter by Professional Journalists: Results of a Newsroom Survey in Germany. In Katrin Weller et al. (eds.). *Twitter and Society,* 345–357. New York, Washington, D.C., Baltimore: Peter Lang.

Olmstead, Kenneth, Amy Mitchell, and Tom Rosenstiel (2011). Navigating News Online: Where People Go, How They Get There and What Lures Them Away. *Journalism*.org, 02/01/2014, http://www.journalism.org/2011/05/09/navigating-news-online/.

Purcell, Kristen et al. (2010). Understanding the Participatory News Consumer. How Internet and Cell Phone Users Have Turned News into a Social Experience. *Journalism.org*, March 1, 02/01/2014, http://www.journalism.org/files/legacy/Participatory_News_Consumer.pdf.

Riepl, Wolfgang (1913 [1972]). *Das Nachrichtenwesen des Altertums. Mit besonderer Rücksicht auf die Römer.* Hildesheim, New York: Georg Olms.

Schmidt, Holger (2011). Trafficlieferanten der Medien: Facebook gewinnt, Google verliert. *faz.net. Netzökonom*, March 29, 02/01/2014, http://blogs.faz.net/netzwirtschaft-blog/2011/03/29/trafficlieferanten-der-medien-facebook-gewinnt-google-verliert-2421/.

Schönhagen, Philomen (1995). *Die Mitarbeit der Leser. Ein erfolgreiches Zeitungskonzept des 19. Jahrhunderts.* München: Reinhard Fischer.

Sehl, Annika (2013). *Partizipativer Journalismus in Tageszeitungen. Eine empirische Analyse zur publizistischen Vielfalt im Lokalen.* Baden-Baden: Nomos.

Singer, Jane B. et al. (eds.) (2011). Participatory Journalism. Guarding Open Gates at Online Newspapers. Oxford: Wiley-Blackwell.

Wahl-Jorgensen, Karin (2007). *Journalists and the Public: Newsroom Culture, Letters to the Editor, and Democracy.* Cresskill, N.J.: Hampton Press.

Wladarsch, Jennifer (2010). *Auf der Spur der Massenmedien in sozialen Onlinenetzwerken. Wie und warum Internetnutzer massenmediale Inhalte in sozialen Onlinenetzwerken nutzen.* Master Thesis: Ludwig-Maximilians-Universität München.

Wladarsch, Jennifer (2014). Journalistische Inhalte in sozialen Onlinenetzwerken: Was Nutzer rezipieren und weiterkommunizieren. In Wiebke Loosen and Marco Dohle (eds.). *Journalismus und (sein) Publikum. Schnittstellen zwischen Journalismusforschung und Rezeptions- und Wirkungsforschung,* 113–130. Wiesbaden: Springer VS.

Changing Technologies, Changing Journalistic Epistemologies: Public Participation, Emotionality and the Challenge to Objectivity

Karin Wahl-Jorgensen

Introduction

This chapter focuses on how new media technologies, in generating new participatory opportunities, might challenge journalistic paradigms, self-understandings, and epistemologies. It takes a particular interest in how long-standing notions of journalistic objectivity might be under fire at a time of shifting relationships between journalists and, as Rosen (2007) put it, "the people formerly known as the audience." The chapter seeks to position the study of journalism with respect to arguments around an "affective turn" across social sciences and humanities disciplines. It argues that despite the historical denigration of emotion in public discourse, ordinary people's personal and emotional stories have always been valued in participatory formats. But what is changing now is that with the increased prevalence of user-generated or amateur content, these emotional stories, told in new and non-objective forms, are no longer physically and spatially marked off as exceptional through their location in specific sections or pages of a newspaper, and hence "ghettoized" or "segregated" (Coward 2013; Wahl-Jorgensen 2008). Instead, they have become part and parcel of news content, challenging epistemologies of journalism.

The chapter focuses on recent historical trajectories in participatory media practices. It examines journalists' and audiences' views on forums for public participation, ranging from letters to the editor to user-generated content. As such, it looks at evolving forms of participatory practice through a particular lens: That of taking an interest in the role of emotion in journalistic story-telling, and how technological change, as it is reshaping conventional boundaries between journalists and their audiences, might destabilize long-standing binary understandings of the relationships between objectivity and emotionality.

Towards an Affective Turn in Journalism Studies?

The argument made in this chapter should be understood in the context of debates over an "affective turn" (Clough and Halley 2007) in humanities and social sciences disciplines, reflecting an increasing interest in how emotional engagements make a difference to social and political life (e.g. Goodwin, Jasper and Polletta 2001; Staiger, Cvetkovich and Reynolds 2010). The place of emotion in public discourse has been hotly debated and contested, and emotionality has historically been denigrated in political thought. Assessments of public discourse in general, and journalism in particular, rely heavily on the normative framework of liberal democratic theory, which has tended to understand emotion as the polar opposite, and enemy of, rationality (e.g. Wahl-Jorgensen 2013a).

The suspicion of emotional engagement and the celebration of rationality inherent in the liberal democratic framework are linked to the circumstances of its birth. Liberal democracy evolved in part in reaction to the longstanding dominance of the church and absolutist sovereigns (Holmes 1995); governance institutions premised on the service of docile subjects. The novel philosophical belief in the rationality of the individual which emerged alongside the rise of science gradually chipped away at the supremacy of sovereigns and the church, in part by insisting on liberty and equality. While the Renaissance and the Enlightenment brought notions of individual reason and the rights of subjects to the forefront of a revolution in political thought, the emergence of democratic thought also always struggled with the relationship between rationality and passion in terms of understanding the parameters of citizen participation. Thinkers such as Hobbes, Rousseau, Locke and Mill, in spelling out the logic of a liberal state that involved rights and participation for citizens, were informed by a concern about the ability of ordinary people to exercise reason and rationality, and about the centrality of public participation as an assurance of the legitimacy of government. As a consequence, the idealization of the rational, dispassionate and informed citizen, coupled with a clear understanding that emotional citizens make for bad subjects, are built into the liberal democratic model. As Marcus, Neuman and MacKuen (2000) summarized it:

> "Western tradition tends to derogate the role of affect in the public sphere. Being emotional about politics is generally associated with psychological distraction, distortion, extremity and unreasonableness. Thus, the conventional view is that our capacity for and willingness to engage in reasoned consideration is too often over-

whelmed by emotion to the detriment of sound political judgment. As a result, theories of democratic practice proclaim the importance of protesting against the dangers of human passion and political faction by building up institutions, rules and procedures—all intended to protect us from our emotional selves." (Marcus et al. 2000, 2)

Along those lines, liberal theory has tended to understand emotion as an inevitable but unfortunate feature of the psychological make-up of citizens, both as individuals and as collectives—as a problem that should be contained and controlled. This view has been particularly important in shaping normative understandings of journalism because media institutions have been understood as central to liberal democratic societies (e.g. McNair 2007), responsible for equipping citizens with the information they need to make rational choices in elections, and for facilitating public debate. Habermas' (1989) account of the role of the media as a key institution in the public sphere—the site for citizen's rational-critical debate on matters of common concern—has been highly influential, and has also exemplified the emphasis on rationality characteristic of modern Western political thought. Thus, journalism's normative role should be viewed through the lens of a historically tense relationship between, on the one hand, long-standing professional ideals of a profession steeped in a liberal democratic framework and, on the other hand, a troubled recognition of the actual inevitability of emotion as a feature shaping everyday life and public discourse as it takes place in and through the media. In journalism studies emotions are often studied in terms of their deviance from ideals of the public sphere (Coward 2013; Pantti et al. 2011). This is evidenced, among other things, in the concern over the sensationalist excesses of tabloid journalism (Sparks and Tulloch 2000) and the "sob sister" style of award-winning journalistic prose (Shapiro 2006). By contrast, any more 'positive' assessments of the role of emotion in journalism have tended to receive less attention. Nonetheless, media and journalism scholars are beginning to take an interest in the role of emotion in the public sphere. This includes some who examine the increasingly close relationship between politics and popular culture (e.g. Jones 2010; Van Zoonen 2005), as well as those who are seeking to develop ideas around an "emotional public sphere" (Lunt and Stenner 2005) and, more concretely, examine the place of emotion in journalistic narratives (Pantti 2010; Peters 2011; Wahl-Jorgensen 2013b; Wahl-Jorgensen 2013c). This also entails attention to the rise of a "subjective journalism" which privileges the personal voice (Coward 2013). Relatedly, journalism scholars are beginning to take heed of the emotional labor carried out by journalists, especially those working in areas associated with conflict and trauma (e.g. Richards and Rees 2011). Further, there is evidence to suggest that journalists themselves are highly aware of the emotional impact of their work on their audiences, and this is increasingly acknowledged by scholars. For example, Zeynep Gürsel's (2010) ethnographic work on photojournalism at an American news magazine relatedly demonstrates that to journalists, the anticipation of audience emotional reactions to stories informs their deliberations over everything from photo selection to layout:

> "Central to the visual appeal is wielding emotions in order to 'bring the story closer' and educate the reader. Novelty of emotion reaction becomes important to photo editors because even with a story the reader might already be assumed to know, a photo-essay is mean to supply an emotional connection that might have been absent in the reader's previous encounters with the story." (Gürsel 2009, 40–41)

The idea of "bringing the story closer" is also central to a set of well-documented journalistic practices around the reporting of distant suffering, as highlighted in the case of humanitarian disasters (e.g. Joye 2009; Joye 2010, Chouliaraki 2006). Here, scholars have pointed out that journalism can cultivate cosmopolitan solidarities through the compassionate reporting of suffering. Nonetheless, such positions, which highlight a potentially interventionist role for journalism, exist in tension with an idea of the journalist as an impartial observer, firmly embedded in the ideal of objectivity.

As a professional aspiration, the notion of objectivity has been central to Western conceptions of journalistic practice, even if it has always been understood as an ideologically inflected contested and counterfactual ideal (Schudson 1982; Ward 2005, 11–13). Journalism gains its cultural authority as truth precisely by drawing on practices associated with objectivity (McNair 1998). This cultural authority is, in turn, underpinned by the belief that it is possible to separate facts from values and the observer from the observed (Ward 2005; McNair 1998). It is premised on a correspondence theory of truth, which assumes that the truth about a particular event or object is constituted by a set of "facts." The more skillfully and objectively a phenomenon is described, the closer the description gets to the truth of it. Such a conception imbues media content and its makers with phenomenal epistemic authority, and is central to arguments about the importance of media professions. Walter Lippmann (1922/1991, 31) famously suggested that to create a correspondence between "the world outside" and the "pictures in our heads," we need a battalion of responsible expert journalists to ensure proper "representation of the unseen facts." Ettema and Glasser (1998) have pointed out that journalism's belief in correspondence theories of truth is captured in the extensive use of metaphors of glass to understand and describe news: When we refer to the news as a "mirror of reality" or a "window on the world," we assume that media content is straightforwardly referential; that there is a one-to-one relationship between the "true facts" of an event, and the reality represented in news stories. Such an understanding points to journalistic accounts as vessels of truthful information. Along those lines, James Carey (1987) referred to journalism in the modern era as a "journalism of information," or a

> "journalism of the expert and the conduit, a journalism of information, fact, objectivity, and publicity. It is a scientific conception of journalism: it assumes an audience to be informed, educated by the journalist and the expert." (Carey 1987, 14)

This "scientific" conception of journalism makes assumptions not only about the subject position of audience members, but also involves a theory of the relationship between journalistic discourses, knowledge and truth. That is to say, the deployment of the "strategic ritual of objectivity" (Tuchman 1972) is also an epistemological practice; a practice which both constructs the truth and, philosophically and ideologically, entails a theory of knowledge.

Epistemology is conventionally understood as the theory of knowledge; the way of explaining how we come to accept particular knowledge claims. Ettema and Glasser (1987), who were among the first scholars to develop the idea of the epistemology of journalism, understood and studied it in terms of how "journalists know what they know." More concretely, they were interested in developing a sociology of epistemology through examining the practices of investigative journalism. They sought to understand what "counts as empirical evidence and how that evidence becomes a justified empirical belief—ergo, a knowledge claim about the empirical world" (Ettema and Glasser 1987, 343), and did this by means of investigating the discursive processes through which knowledge claims are justified in journalism.

This chapter, however, understands the epistemology of journalism more broadly in terms of the "*rules, routines* and *institutionalized procedures* that operate within a social setting and decide the form of the knowledge produced and the *knowledge claims* expressed (or implied)" (Ekström 2002, 260, emphasis in original). This is particularly important to consider in the light of journalism's epistemological position as the "primary sense-making practice of modernity" (Hartley 1996, 32–34). That is to say, the knowledge claims of journalism have broader ideological consequences, but are also shaped by sociological forces and prevailing power relations. Here, I am drawing on recent work investigating how knowledge claims are produced by specific forms of journalism, including blogs (Matheson 2004) and broadcast journalism (Ekström 2002). This approach entails an understanding of how journalists' processes of justification involve the construction of narratives that operate within conventions established by institutional forms of knowing, circumscribed by power relations. As Matheson (2004) described it, drawing on a Foucauldian analysis of the relationship between knowledge and power:

> "Conventions of newswriting do not simply chronicle the world but […] constitute certain claims to knowledge about such matters as the audiences for news texts, the position of journalists in that world and the relationship between audience and journalist. […] Journalists adhere to these conventions in order to be able to make the kinds of authoritative statements about events and individuals which we are accustomed to hear from them. News discourse can be seen as a particular instance of the more general 'will to truth' which motivates and constrains institutional forms of knowing in modern society." (Matheson 2004, 445)

In investigating the epistemology of blogging, Matheson (2004) took a closer look at how the British newspaper, *The Guardian*, responded to the introduction of the new form in its own hosted blogs. He demonstrated that they were characterized by a distinctive way of knowing premised on the "establishment of a different interpersonal relation, of a different authority and of a journalism focused upon connection rather than fact" (Matheson 2004, 453). To Matheson, the writing represented a "more 'raw,' less 'cooked,' source of information, allowing users to participate more in constructing knowledge about events in the world" (Matheson 2004, 455). The suggestion that technological change has a bearing on the ways of knowing in journalism and that it may contribute to shifting the subject positions and interpretations of the audience is crucial to the argument made in this chapter. Here, I take an interest in understanding how journalistic ways of knowing have been shaped both by historical processes of professionalization and de-professionalization, and by technological change. In particular, I argue that if the distinctions between "fact" and "personal opinion," and between rational information and emotional story-telling have long been foundational to conventional journalism, these distinctions are rapidly being undermined in the digital era with the increasing prevalence of user-generated or amateur content, with profound epistemological and, needless to say, practical consequences. I want to consider these in the context of shifts in forms of mediated public participation, suggesting that we can trace over time "the rules, routines and institutionalized procedures that operate within a social setting and decide the form of the knowledge produced and the knowledge claims expressed (or implied)" (Ekström 2002, 260).

Here, I am drawing on the received account of journalism history, suggesting if the first (and non-professional) newspapers made little distinction between "news" and "opinion," professional journalism came to be understood as a "fact-centered discursive practice" premised on ideals of objectivity (Chalaby 1999). This understanding, however, is now under fire from the increasing importance of user-generated or amateur content, which is swiftly ushering in new ways of knowing, or new ways of determining what counts as truth within the interpretive communities of journalism, often grounded in more emotional, subjective and partial story-telling practices (Fish 1980; Zelizer 1993).

Changing Technologies, Changing Epistemologies of Mediated Public Participation

Early print publications—by most[1] measures the first mass media—made little distinction between opinion and news content, and, correspondingly, between opinion pieces in the form of letters to the editor and factual reports on current events. In the prominent account of Jürgen Habermas (1989), for example, the first newspapers emerged as the organic continuation of private newsletters. Before the advent of the steam train and the telegraph, printers relied on news arriving by postal coach or ship, and stories published in the early British and American press were often several months old. In this information-poor environment, printers and editors welcomed any publishable material. Philadelphia printer Andrew Bradford, writing at the beginning of the 18th century, justified his printing of a personal letter in terms of the lack of other suitable news: "Having but few remarkable occurrences to fill up our paper at present, we believe it will not be unacceptable to our readers, to incert [sic!] the following letter from a Gentleman to his Friend upon the

1 Some examples used in this section have been discussed in a different context in the following pubication: Wahl-Jorgensen, Karin (2014). From Letters to Tweeters: Media Communities of Opinion. In Martin Conboy and John Steele (eds). *Routledge Companion to British Media History.* London: Routledge.

loss of his only Daughter" (Hart 1970, 113). This dynamic encouraged the inclusion of correspondence from members of the public, but also reflected the fact that the category of the professional journalist, separate from the printer who simply compiled incoming information, had yet to be born.

The separation of news and opinion, and therefore of stories from letters, was integral to the slow but steady professionalization of journalism. One indication of this separation can be seen in the demarcation of letters to the editor as a distinctive genre and forum within the newspaper, separated from content produced by professionals. For example, *The New York Times* published its first letter to the editor five days after its first issue came out, on September 18, 1851 (Rosenthal 1971, 135). The paper sought to make its letters section "a forum for the consideration of all questions of public importance, and to that end to invite intelligent discussion from all shades of opinion" (Seigel 1972, 3–4). This is illustrative of the ways in which, as a forum for public participation, the letters section has been associated with public sphere ideals associated with "intelligent discussion" and privileging an idea of rational-critical debate consistent with a liberal democratic framework. During the lifetime of pre-convergence newspapers, letters to the editor served as the primary site for the inclusion of audience perspectives. Most print newspapers have a dedicated letters section as part their opinion pages, often situated on their op-ed pages and, as such, clearly delineated as a space dedicated to a distinctive opinion-based expression which relies on a different epistemology from the fact-centered discursive practices of "hard news" journalism[2] (Wahl-Jorgensen 2008). Nonetheless, despite the fact that the letters section, as a key forum for public debate, has relied on a normative ideal of rational-critical debate, the forms of expression that journalists actually favor in letters have often been at odds with this ideal.

In my own research on journalistic practices associated with letters to the editor, carried out in the San Francisco Bay Area, and consisting of ethnographic research in one newsroom and interviews with 24 newsworkers across newspaper types involved with selecting and editing letters, the tension between ideals and practice was apparent: On the one hand, newsworkers spoke glowingly of the letters section as an ideal "open forum" for rational discussion among members of their communities (Wahl-Jorgensen 2007). On the other hand, they despaired of the frequent contributions from "nut

2 With the exception of some local newspapers, where letters to the editor have been published alongside articles written by journalists (e.g. Wahl-Jorgensen 2007).

cases" or "crazies" who hold irrational, extreme and polarized positions and are unwilling to engage in civil debate with other letter writers. Along those lines, they often characterized letter writers in terms of their irrationality, deploying the "idiom of insanity" (Wahl-Jorgensen 2002a). They were constantly thinking about how to take steps to improve the quality of debate. One editorial page staff member at the newspaper where I carried out my ethnographic research—a daily flagship newspaper in a small local chain—offered me an example of his efforts at encouraging contributions critical of the frequent outpourings against then-President Bill Clinton and his wife:

> "There was a group called the Hillary Club in Orinda, and they're Hillary Clinton fans. And, they were just appalled by something in the paper and they were pretty sure that we hated her, which we don't. We talked for a long time and part of her accusation was that we ran these crazy, Clinton-haters, that say the same thing all the time. I said yes, we do, but they write, and you don't."

His solution to the perceived problem of insanity in letters, then, lay in a trickle-down theory of public discourse: By introducing a rational or even elite discourse as a counterpoint to the discourse of the "crazies," the public sphere would be transformed into a safe neighborhood in which to be heard and seen. This view, and others closely associated with it, privileged a rational and civil form of public debate on important issues of common concern, over one which was emotional, subjective, polarized and polarizing. Yet when the newsworkers I interviewed expanded on which types of letters they preferred, it became clear that in fact, their preference is for personalized and deeply emotional stories. One editor for a small weekly paper in an affluent community struggling to come to terms with issues of urban development and growth, explained this preference for letters in the form of such personalized stories from local residents in terms which contrasted them with the discourse of lobbying organizations which were using "political rhetoric." His account is worthwhile quoting at length, as it captures prevailing views of preferred forms of public debate expressed more broadly by newsworkers selecting and editing letters:

> "The resident tends to speak plainly and from the heart, more than some of the political rhetoric of some of these organizations. You see the same kind of rhetoric from some of these groups. It might be very heartfelt, but a lot of times you'll see things like and we appreciate all the many volunteers that can make this organization the organization that it is that it can indeed be all it can be. We've made many strides and much progress in achieving our goals. And then they don't say what those goals are, or specifically how they've made strides. It's almost like they're out to at-

tract more membership than do good deeds. […] Individuals have something more specific to say, and they're usually more honest about saying it. […] Here's Olinger's [anti-development letter writer] letter in the paper where he says that "from where I stand, there's no rural anymore," whereas a group like the Civic Association is going to be lambasting town government for its time-consuming process of granting applications to deserving home-owners who deserve to build their dream-home in the hills. […] So you're not getting the truth and the whole truth out of them as an organization. You are getting the truth from this individual, in this letter, 'cause this is how he's feeling, and that's why he's written this letter. He's not mincing words, he's not shielding his true intentions in a bunch of mumbojumbo […] Olinger […] how are you going to challenge him? Are you going to say "I don't agree that there's no rural anymore"? You can't challenge his opinion."

Like many other editors, this one celebrates the "heartfelt" discourse of the regular citizen who openly talks about his feelings, juxtaposing it with the stale, rehearsed "boilerplate" rhetoric of the activist groups, a discourse that is "kind of manufactured, coming off of an assembly line," or "just like an ad," as other editors put it. This position emphasizes the sincerity, authenticity, and truth, and concreteness of the regular citizen and their personalized storytelling precisely because of its emotional integrity. This, in turn, represents a challenge to the epistemological assumptions of conventional journalism, where the authority of "truth" in journalistic discourses is associated with ideas of objectivity and professionalism. However, in this context, such an understanding of truthfulness—as represented by emotional and personalized story-telling—is confined to a specific section of the newspaper. It is thus "segregated" or "ghettoized" away from the professional discourses which are underpinned by the ideal of objectivity. Moreover, the letters section has always served as a *constructed* community of mediated public opinion, where decisions about which letters to include and which to reject have been made on the basis of distinctive and institutionalized rules of selection, implemented by newsworkers who have served as gatekeepers to guarantee the quality of contributions (Wahl-Jorgensen 2002b).

Even if letters were, for a long time, the most prominent site for mediated public participation, other traditional mass media—particular radio and television—have given birth to a range of forums for mediated discussion. Audience participation in radio genres has been around as long as the medium itself, and has consistently offered a space for the representation of the public, claiming to speak for "the people" (Loviglio 2002). For instance, the network radio program *Vox Pop*, broadcast between 1932 and 1948, engaged in an unending search for the voice of the American people by interviewing individuals in the streets, with the stated intention of "posing questions of 'spectacular unimportance'" (Loviglio 2002, 91). Community and shortwave radio stations have provided an important way for otherwise disenfranchised groups to gain a foothold in the public sphere (Fairchild 2001, 89; Riismandel 2002). Talk radio has been a key venue for "populist deliberation," through which citizens may bypass the mainstream media to make their opinions heard and hold politicians to account (Page and Tannenbaum 1996; Thornborrow and Fitzgerald 2013), even if many observers have raised concerns over the incivility and intolerance often characterizing the genre, as well as for the its domination by people with extreme and polarized political positions (Sobieraj and Berry 2011)—mirroring the concerns expressed by journalists working with letters to the editor over the "irrationality" of writers.

Platforms for opinion expression in conventional print and broadcast media—including letters to the editor, television and radio talk shows, and audience participation programs—have allowed editors, journalists and program makers to shape the tone and content of mediated public debate. They have functioned as an integral part of media content—whether as a section in the newspaper or as incorporated into programming in the form of television news vox pops or programming premised on audience participation. This mode of incorporation into conventional media operation and professional practices has, however, been challenged by technological change. This, in turn, complicates the relationship between access, mediation and professional intervention in and contribution to public debate.

The emergence of the internet and, subsequently, convergent forms of news content enabling greater interactivity in a proliferation of forums, genres and forms—ranging from blogs, comments and user-generated content to social media—has had profound consequences. First of all, the technological affordances of the internet have enabled greater interactivity (e.g. Kammer 2013). Starting with the earliest experiments in the 1990s and early 2000s, media organizations enabled users to comment on online stories, and the introduction of blogs also enabled further instantaneous dialogue, for the first time generating communities of opinion that could respond in real time to unfolding news events (Steensen 2011). The internet was welcomed with much fanfare by observers who saw it as an opportunity to "produce virtual public spheres" (Papacharissi 2002) and hence revolutionize mediated public participation. New technologies have certainly contributed to destabilizing power relations in the public sphere, particularly with respect to the production and distribution of content and opinion. To many, these trends began with the emergence of citizen journalism and user-generated content. User-generated content—or views, images and videos contributed by members of the public—first gained prominence after the 2004 Indian Ocean tsunami, where eyewitnesses were able to film the disaster as it unfolded, providing news organizations with unprecedented immediacy in their coverage (Allan 2009). In the UK, the 7/7 bombings in July 2005 represented the watershed for "accidental journalism" carried out by ordinary citizens (Boaden 2008; Allan 2013). The use of audience materials in the context of major breaking news events, particularly natural disasters, was focused on information provision, rather than the sharing of opinion—even if such "citizen witnessing" has always taken place from a particular subject position which implies, at the very least, an investment and an interest in current affairs (Allan 2013; Andén-Papadopoulos and Pantti 2011). News organizations' heady embrace of opinion-based user-generated content over the next few years heralded the increasing prominence and centrality ascribed to such audience participation, which was seen as a useful venue for generating public debate on major news stories in ways that might both enhance public participation, bolster relationships with audiences, and generate sources for follow-ups and future stories. The increasing use of audience materials in the conventional media worked in tandem with the creation of new platforms and forums where audiences could participate by contributing news and opinion—leading to the rise of "citizen journalism" (e.g. Allan and Thorsen 2009). In a keynote speech and subsequent blog post on the phenomenon published in 2008, the BBC's then-director of news, Helen Boaden, framed the growing awareness of and resources shifted in the direction of citizen journalism within the corporation in terms of audience relations:

> "The biggest challenge for us is about our relationship to the people who matter most—our audiences. It's about capturing and keeping their hearts and minds. And for audiences who want to join in, that means including them in the process of making the news. Our journalism is now fully embracing the experiences of our audiences, sharing their stories, using their knowledge and hosting their opinions; we're acting as a conduit between different parts of our audience; and we're being more open and transparent than we have ever been. […] [W]ith blogs in particular—but also podcasts and videoblogs—the ability of the public to express opinion in public has exploded—especially in the USA—and they no longer need to be "hosted" by broadcasters." (Boaden 2008)

Boaden's position is typical of journalists' complex orientation towards the proliferation of mediated forums for sharing and creating debate, content and opinion: The promise of new platforms as an unprecedented opportunity to connect with audiences has often been a key driver for both commercial and public service media. At the same time, for news media hosting—and therefore taking responsibility for—such spaces, they also require extensive vigilance. Concerns about the monitoring, moderation and quality control of online forums have remained central for news organizations and have, over time, led to more careful constraints and limitations on spaces for discussion, with the closing down of discussion boards and comments originally hosted by broadcasters and newspapers (Wahl-Jorgensen, Wardle and Williams 2010).[3] Such concerns highlight a contemporary twist on an apparent paradox that has always plagued sites for public debate and opinion formation: They are viewed as central to democratic practice and the formation of a public sphere for discussion of matters of common concern, and as such as a key responsibility of media organizations. But their management requires extensive resources, take away power from news organizations; and individuals participating in discussion through such venues are not necessarily reflective of the profile or views of a broader public. The shift also challenges conventional power relations of the public sphere, where participation in the form of opinion expression is no longer the preserve of mainstream media. Instead, participants have been granted a greater autonomy over the production and distribution of opinion. Media organizations, which have tended to function as the "gatekeepers," have now become "gatewatchers" or curators, sorting through and publicizing information available elsewhere on the internet (Bruns 2005, 2). Perhaps most importantly, the increasing role of user-generated content represents a shift from the situation where audience contributions are consistently and neatly parceled off in specific sections, and towards one where they are, at least on occasion, broadcast or published alongside content provided by professional journalists. This is significant, because it means that the clashing and fundamentally incompatible epistemologies of conventional "objective" journalism and "emotional" audience content now sit alongside each other, rather than the former being

3 For example, the BBC closed down a range of sports and music-related discussion boards following the Graf Report into the corporation's online activities in 2004 (http://news.bbc.co.uk/1/hi/entertainment/3866355.stm, 08/29/2013). This move reflected a broader view of such discussion boards as difficult to moderate and relatively peripheral to the public service mission of the corporation.

privileged by the hierarchies of news content. It coincides with a trend of the rise of "subjective journalism" which privileges personal voice, alongside a broader democratization of opinion—both of which deal a blow to conventional understandings of news journalism as the voice of authority (Coward 2013, 116–117).

The celebration of the perceived authenticity—and the closely related emotionally engaging nature of ordinary people's contribution—discussed in the case of letters to the editor also carries over journalists' and audiences' views of user-generated content, as demonstrated by our research on user-generated content in the BBC, carried out through interviews, focus groups and ethnographic observation in 2007 and 2008. We found that both audience members and journalists valued user-generated in the form of amateur footage on the basis of its perceived authenticity and emotional content, deliberately juxtaposing it to the more objective and distanced, and hence less engaging, professional discourse of journalists. One focus group, for example, discussed the empathy invited in viewers by news packages which include footage of extraordinary events filmed by ordinary people—in this case a woman filming her living room being flooded. They contrasted the poignancy of this footage with the cold detachment they imagined a professional journalist or camera person would bring to the coverage of the same story:

June: "Well I think that a professional person couldn't feel like that woman, because it's her home, so she's giving a more accurate […] she's bringing it home to you more than if it was a professional person in her house."
Juliet: "That's really important actually, I never thought of it like that. The person behind the camera is feeling all of those things, and you are seeing it literally through her eyes […] what's important to them, like."
Christopher: "It's not as emotive by any stretch of the imagination if it's professional footage."

These audience members valued user-generated content over ostensibly "objective" news stories produced by professionals because it provided "emotive" information about what the person behind the camera was "feeling." Such a position represents a direct challenge to the journalistic paradigm of objectivity, and appears to draw on an epistemological vocabulary which equates truth with authenticity, emotional integrity and immediacy. This is a vocabulary which both mirrors and draws on assessments of reality TV genres, which are playing an increasingly central role in programming (e.g. Hill 2004). This view, which came up in many of our focus groups, has significant implications, suggesting that conventional journalistic ways of knowing, based on ideals of objective reporting, may not in fact be seen by audiences as encapsulating the "truth" as much as more emotive forms of storytelling.

Many of the newsworkers involved with user-generated content drew on similar tropes in explaining the forms of participation from "ordinary people" that they found most compelling. We interviewed the Capture Wales team—a group of media producers at BBC Wales that worked with ordinary people to tell their stories through the use of digital technologies, including photography and video. Carwyn Evans, a member of this team, explained his preference for stories that embodied "passion" through personalized storytelling, juxtaposing it with the "depersonalizing" content of mainstream media:

> "And for me it's all about people realizing their potential and exceeding what they thought they could do. But obviously though the thing that makes a good story for me is the fact that their passion comes across. It doesn't have to be sensational, it's the simple [stories], their favorite pair of shoes or talking about how much they loved their parents, simple things in life. [...] My ears prick up when I hear somebody, even if it's on the radio, when I hear somebody talking about something they're passionate about, it really touches me. That's what's great about these kinds of forms. They are totally different to what you get on conventional TV and national radio. I mean, on any reporting, take any, take what's going on in Iraq for instance. You know, what do you hear about, you don't hear about the past or you don't hear about the real excellent people, you don't hear about what's it like to get up in the morning and not have water or ... You don't get any sense ... [Mainstream media] depersonalizes."

The 'Evans' view of the Capture Wales Team, though distinctive in its rejection of the epistemologies of mainstream media, was representative of broader arguments among newsworkers at the BBC about the value of user-generated content in providing authentic, embodied and passionate stories from "ordinary people." However, while many of the journalists we interviewed valued audience content for much the same reasons as members of the public, they were also keen to stress the continued need for journalistic skills, upholding the boundary between journalism professionals and audience members as amateurs. Along those lines, a BBC course on user-generated content reflected this broader view, and was titled "Have They Got News For US?" (Williams et al. 2011). When we interviewed then-Radio Editor Peter Rippon about the public service broadcaster's use of audience materials, he suggested that: "If we apply our journalistic skills, you can just mould these things into really good pieces of journalism." This statement was typical of the journalistic position on user-generated content, which argued unequivocally that professional journalistic skills were required to turn amateur, unpolished contributions into useable content. Ultimately, such boundary work (Schudson and Anderson 2009) turns the audience into a provider of supplementary emotional and emotive rather than rational material (Williams et al. 2011) to protect journalists' privileged status as producers of knowledge and truth. It seeks to defend news organizations against the encroaching breakdown of the boundary between objective and professional ways of knowing that have for so long been privileged, and the emotional, embodied and amateur storytelling which appears to be increasingly prominent.

Conclusion: Towards Postmodern Journalism?

This chapter has argued that technological change is affecting not only journalistic and audience perceptions of what is valued in journalism, but also the profession and its practices and epistemologies. As we examine changes in the technologies of journalism, then, we have to ask ourselves broader questions not merely about the consequences for the profession but also about the nature of public discourse. In particular, the chapter has focused on the increasing prevalence of user-generated content as an integral part of the provision of legacy news organizations. Both in a very physical but also in a conceptual sense, the distinction between journalistic and audience labor has blurred. The epistemology of conventional "fact-centered" journalism, associated with the ideal of objectivity, may no longer (if indeed it ever has been) bevalued by audiences. The shifts described here may over time lead to a more permanent change in the presentation and styles of news, including the increasing prominence and naturalization of what Coward (2013) has referred to as "subjective journalism." This trend, the chapter has suggested, introduces a more emotionalized form of public discourse—and hence a way of knowing – which is described by both audience members and newsworkers in juxtaposition to ideals of objectivity.

The blurring of conventional boundaries introduced by technological change could be understood as postmodern form of journalism. I am using this phrase to gesture in the direction of a form of journalism which has destabilized conventional (a) physical, stylistic and genre distinctions; (b) differentiations between amateur and professional content, and (c) epistemological distinctions around the truth value of objective versus emotional content. As we have seen, this development compels journalists to engage in "boundary work" to defend the integrity of their profession. However, such boundary work might well be all for nought, and instead of dismissing the changes caused by postmodern journalism, we would do well to study them in detail and understand their consequences.

Works Cited

Allan, Stuart (2009). Histories of Citizen Journalism. In Stuart Allan and Einar Thorsen (eds.). *Citizen Journalism: Global Perspectives*, 17–32. New York: Peter Lang.

Allan, Stuart (2013). *Citizen Witnessing*. Cambridge: Polity.

Allan, Stuart, and Einar Thorsen (2009). Introduction. In Stuart Allan and Einar Thorsen (eds.). *Citizen Journalism: Global Perspectives*, 1–16. New York: Peter Lang.

Andén-Papadopoulos, Kari, and Mervi Pantti (eds.) (2011). *Amateur Images and Global News*. Bristol: Intellect Books.

Boaden, Helen (2008). The Role of Citizen Journalism in Modern Democracy. *BBC News—Editors' Blog*, November 13, 09/04/2013, http://www.bbc.co.uk/blogs/theeditors/2008/11/the_role_of_citizen_journalism.html.

Bruns, Axel (2005). *Gatewatching: Collaborative Online News Production*. New York: Peter Lang.

Carey, James (1987). The Press and Public Discourse. *The Center Magazine*, 20, 4–32.

Chalaby, Jean K. (1998). *The Invention of Journalism*. Basingstoke: Macmillan.

Chouliaraki, Lilie (2006). *The Spectatorship of Suffering*. London: Sage.

Clough, Patricia T., and Jean Halley (eds.) (2007). *The Affective Turn: Theorizing the Social*. Durham, N.C.: Duke University Press.

Coward, Rosalind (2013). *Speaking Personally: The Rise of Subjective and Confessional Journalism*. Houndsmills, Basingstoke: Palgrave Macmillan.

Ekström, Mats (2002). Epistemologies of TV Journalism: A Theoretical Framework. *Journalism*, 3 (3), 259–282.

Ettema, James S., and Theodore L. Glasser (1987). On the Epistemology of Investigative Journalism. In Michael Gurevitch and Mark R. Levy (eds.). *Mass Communication Review Yearbook*, 6, 338–361. Newbury Park, Calif.: Sage.

Ettema, James, and Theodore L. Glasser (1998). *Custodians of Conscience: Investigative Journalism and Public Virtue*. New York: Columbia University Press.

Fairchild, Charles (2001). *Community Radio and Public Culture.* Cresskill, New Jersey: Hampton Press.

Fish, Stanley (1980). *Is There a Text in This Class: The Authority of Interpretive Communities.* Cambridge, Mass.: Harvard University Press.

Goodwin, Jeff, James M. Jasper, and Francesca Polletta (eds.) (2001). *Passionate Politics: Emotions and Social Movements.* Chicago: University of Chicago Press.

Gürsel, Zeynep D. (2009). U. S. Newsworld: The Rule of Text and Everyday Practices of Editing the World. In S. Elizabeth Bird (ed.). *The Anthropology of News and Journalism,* 35–53. Bloomington, Indiana: Indiana University Press.

Habermas, Jürgen (1989). *The Structural Transformation of the Public Sphere.* Cambridge, Mass.: MIT Press.

Hartley, John (1996). *Popular Reality.* London: Arnold.

Hill, Annette. (2004). *Reality TV: Audiences and Popular Factual Television.* New York, London: Routledge.

Holmes, Stephen (1995). *Passions and Constraint: On the Theory of Liberal Democracy.* Chicago: University of Chicago Press.

Jones, Jeffrey P. (2005). *Entertaining Politics: New Political Television and Civic Culture.* Lanham, Md.: Rowman & Littlefield.

Joye, Stijn (2009). The Hierarchy of Global Suffering: A Critical Discourse Analysis of Television News Reporting on Foreign Natural Disasters. *The Journal of International Communication,* 15 (2), 45–61.

Joye, Stijn (2010). News Discourses on Distant Suffering: A Critical Discourse Analysis of the 2003 SARS Outbreak. *Discourse & Society,* 21 (5), 586–601.

Kammer, Aske (2013). *News on the Web: Instantaneity, Multimodality, Interactivity, and Hypertextuality on Danish News Websites.* Ph.D thesis: University of Copenhagen.

Lippmann, Walter (1991 [1922]). *Public Opinion.* New Brunswick, N.J.: Transaction Publishers.

Loviglio, Jason (2002). Vox Pop: Network Radio and the Voice of the People. In Michelle Hilmes and Jason Loviglio (eds.). *Radio Reader: Essays in the Cultural History of Radio,* 89–112. New York, London: Routledge.

Lunt, Peter (2009). Television, Public Participation, and Public Service: From Value Consensus to the Politics of Identity. *The Annals of the American Academy of Political and Social Science,* 625 (1), 128–138.

Lunt, Peter, and Paul Stenner (2005). The Jerry Springer Show as an Emotional Public Sphere. *Media, Culture & Society,* 27 (1), 59–81.

Marcus, George E., W. Russell Neuman, and Michael MacKuen (2000). *Affective Intelligence and Political Judgment.* Chicago: University of Chicago Press.

Mascheroni, Giovanna (2012). Online Participation: New Forms of Civic and Political Engagement or Just New Opportunities for Networked Individualism. In Brian D. Loader and Dan Mercea (eds.). *Social Media and Democracy: Innovations in Participatory Politics,* 207–223. London, New York: Routledge.

Matheson, Donald (2004). Weblogs and the Epistemology of the News: Some Trends in Online Journalism. *New Media & Society*, 6 (4), 443–468.
McNair, Brian (2007). *An Introduction to Political Communication*. 4th ed. London, New York: Routledge.
McNair, Brian (1998). The Sociology of Journalism. Oxford: Oxford University Press.
Mutz, Diana (2006). *Hearing the Other Side: Deliberative versus Participatory Democracy.* Cambridge, New York: Cambridge University Press.
Pantti, Mervi (2010). The Value of Emotion: An Examination of Television Journalists' Notions on Emotionality. *European Journal of Communication*, 25 (2), 168–181.
Pantti, Mervi, Karin Wahl-Jorgensen, and Simon Cottle (2012). *Disasters and the Media.* London, New York: Peter Lang.
Papacharissi, Zizi (2002). The Virtual Sphere: The Internet as a Public Sphere. *New Media & Society*, 4 (1), 9–27.
Peters, C. (2011). Emotion Aside or Emotional Side? Crafting an 'Experience of Involvement' in the News. *Journalism*, 12 (3), 297–316.
Richards, Barry, and Gareth Rees (2011). The Management of Emotion in British Journalism. *Media, Culture & Society*, 33 (6), 851–867.
Riismandel, Paul (2002). Radio by and for the Public: The Death and Resurrection of Low-power Radio. In Michelle Hilmes and Jason Loviglio (eds.). *Radio Reader: Essays in the Cultural History of Radio,* 423–451. New York, London: Routledge.
Rosen, Jay (2006). The People Formerly Known as the Audience. *PressThink: Ghost of Democracy in the Media Machine*, June 27, 01/04/2013, http://journalism.nyu.edu/pubzone/weblogs/pressthink/2006/06/27/ppl_frmr.html.
Rosenthal, Irving (1971). Who Writes the 'Letters to the Editor'? In Allen Kirschner and Linda Kirschner (eds.). *Journalism: Readings in Mass Media*, 135–142. New York: Odyssey Press.
Schudson, Michael (1978). *Discovering the News*. New York: Basic Books.
Seigel, Kalman (1972). *Talking Back to the* New York Times*: Letters to the Editor, 1951–1971,* New York: Quadrangle Books.
Shapiro, Stephanie (2006). Return of the Sob Sisters. *American Journalism Review*, 12/16/2013, http://www.ajr.org/article.asp?id=4117m.
Sobieraj, Sarah, and Jeffrey M. Berry (2011). From Incivility to Outrage: Political Discourse in Blogs, Talk Radio, and Cable News. *Political Communication*, 28 (1), 19–41.
Sparks, Colin S., and John Tulloch (eds.) (2000). *Tabloid Tales: Global Debates over Media Standards*. Lanham, Md.: Rowman & Littlefield.
Staiger, Janet, Ann Cvetkovich, and Ann Reynolds (eds.) (2010). *Political Emotions: New Agendas in Communication*. London, New York: Routledge.
Steensen, Steen (2011). Online Journalism and the Promises of New Technology: A Critical Review and Look Ahead. *Journalism Studies*, 12 (3), 311–327.

Tuchman, Gaye (1972). Objectivity as Strategic Ritual: An Examination of Newsmen's Notions of Objectivity. *American Journal of Sociology*, 77, 660–679.

Van Zoonen, Liesbet (2005). *Entertaining the Citizen: When Politics and Popular Culture Converge*. Lanham, Md.: Rowman & Littlefield.

Wahl-Jorgensen, Karin (2002a). The Construction of the Public in Letters to the Editor: Deliberative Democracy and the Idiom of Insanity. *Journalism*, 3 (2), 171–293.

Wahl-Jorgensen, Karin (2002b). Understanding the Conditions for Public Discourse: Four Rules for Selecting Letters-to-the-editor. *Journalism Studies,* 3, 69–81.

Wahl-Jorgensen, Karin (2007). *Journalists and the Public: Newsroom Culture, Letters to the Editor, and Democracy*. Creskill, N.J.: Hampton Press.

Wahl-Jorgensen, Karin (2008). Op-ed pages. In Bob Franklin (ed.). *Pulling Newspapers Apart*, 70–79. London: Routledge.

Wahl-Jorgensen, Karin (2013a). Future Directions for Political Communication Scholarship: Considering Emotion in Mediated Public Participation. In Kelly A. Gates (ed.). *Media Studies Futures,* 455–478. Oxford: Blackwell.

Wahl-Jorgensen, Karin (2013b). The Strategic Ritual of Emotionality: A Case Study of Pulitzer Prize-winning Articles. *Journalism*, 14 (1), 129–145.

Wahl-Jorgensen, Karin (2013c). Subjectivity and Story-telling in Journalism: Examining Expressions of Affect, Judgement and Appreciation in Pulitzer Prize-winning Stories. *Journalism Studies,* 14 (3), 305–320.

Wahl-Jorgensen, Karin (forthcoming, 2015). *Emotions, Media and Politics*. London, New York: Polity Press.

Wahl-Jorgensen, Karin, Andrew Williams, and Claire Wardle (2010). Audience Views on User-generated Content: Exploring the Value of News from the Bottom Up. *Northern Lights*, 8, 177–194.

Ward, Stephen J. A. (2005). *Invention of Journalism Ethics: The Path to Objectivity and Beyond.* Montréal, Kingston: McGill-Queen's University Press.

Zelizer, Barbie (1993). Journalists as Interpretive Communities. *Critical Studies in Media Communication*, 10 (3), 219–237.

Notes on Contributors

Kevin G. Barnhurst is Professor of Communication in the Digital Era at the University of Leeds. He has been a guest professor or visiting scholar at universities in Denmark, Finland, and Spain and was Distinguished Fulbright Chair in Italy, Shorenstein Fellow at Harvard University, Research Fellow at Columbia University in New York City, and Senior Fulbright Scholar in Peru. He is also Professor Emeritus of Communication, University of Illinois at Chicago, and Senior Visiting Professor of Visual Communication at IE University, Madrid, Spain. His publications include *Media queered: visibility and its discontents* (Peter Lang, 2007). With John Nerone he authored *The Form of News. A History* (Guilford Press, 2001). His more than 100 scholarly publications include articles in *the Journal of Communication, Media, Culture & Society*, and many other journals, book chapters, and shorter pieces, in English, Italian, Portuguese, and Spanish.

Tobias Eberwein is currently Professor of Communication with a focus on structures of public communication at Dresden University of Technology. His research encompasses topics such as online and print journalism, international journalism, media ethics and media accountability, as well as methods of communication and media research. His most recent publications include a monograph on literary journalism in Germany (*Literarischer Journalismus. Theorie – Traditionen – Gegenwart*. Herbert von Halem Verlag, 2013) and the anthology *Journalists and Media Accountability. A Study of News People in the Digital Age* (co-edited with Susanne Fengler, Gianpietro Mazzoleni, Colin Porlezza and Stephan Russ-Mohl; Peter Lang, 2014). Tobias Eberwein is the editor of *Journalistik*, a learned journal for German journalism studies, and of *Aviso*, the news service of the German Communication Association.

Andreas Fickers is Professor of Contemporary and Digital History at Luxembourg University. His main research interests are transnational media history and cultural history of media technologies as well as epistemological and methodological questions of digital history. He is coordinator of the European Television History Network and editor in chief of *VIEW Journal of European Television History and Culture* (www.viewjournal.eu). Recent publications include *A European Television History* (edited with Jonathan Bignell; Blackwell-Wiley, 2008), *Materializing Europe. Transnational Infrastructures and the Project of Europe* (edited with Alexander Badenoch. Palgrave-Macmillan, 2010); *Transnational Television History: A Comparative Approach* (edited with Catherine Johnson; Routledge, 2012); and *Airy Curtains in the European Ether: Broadcasting and the Cold War* (edited with Alexander Badenoch and Christian Henrich-Franke; Nomos Verlag, 2013).

Rita Gudermann is a historian with working experience as a researcher and lecturer at the Free University of Berlin, the Humboldt-University Berlin, the Berlin-Brandenburgische Akademie der Wissenschaften and the Leibniz-Institut für Regionalentwicklung und Strukturplanung in Erkner since 1993. From 2001 to 2004, she worked as a freelance journalist (ZEIT, WDR, Deutschlandfunk) and exhibition organizer (Industrie- und Filmmuseum Bitterfeld). She is researcher in the DFG-Project "Ware Information" at FU Berlin. Her publications range from ecological and economical questions of drainage and irrigation in 19th century Prussia to agrarian statistics (1750–1880) and the history of the "Sarotti Mohr" as an ambivalent advertising character. She is currently working on the commercialization of content as a virtual asset in the 19th and 20th centuries.

Jens Jäger is Heisenberg-fellow at the University of Cologne and currently Professor for Modern History at the University of Heidelberg. He has studied Modern History at the University of Hamburg where he received his Dr. phil in 1995. He is specialized in all questions pertaining to images as sources. He has written on the relationship between photography, history and modern society in the 19th and 20th centuries. His fields of research include Modern German and European History with a focus on the history of police, criminality, international policing, colonial history and the history of media. Recent publications include *Fotografie als historisches Dokument* (in: Fotogeschichte 32:124, 2012, 13–18), *Fotografie und Geschichte* (Campus, 2009) and *Heimat in Africa. The reconfiguration of the German Nation* (in: German History 27:4, 2009, 467–489).

John Nerone is Professor Emeritus of Communications at the University of Illinois at Urbana-Champaign. He is the author, co-author, or editor of five books and numerous articles, including *Violence Against the Press: Policing the Public Sphere in U.S. History* (Oxford University Press, 1994), *Last Rights: Revisiting Four Theories of the Press* (University of Illinois Press, 1995), *The Form of News: A History* (with Kevin Barnhurst; Guilford Press, 2001), which won the Suzanne Langer Award of the Media Ecology Association and the International Communication Association's Book of the Year Award, and *Media History and the Foundations of Media Studies* (Wiley-Blackwell, 2013). He also co-edits (with Robert W. McChesney) a book series on the History of Communication.

Christoph Neuberger is Professor at the Department of Communication Science and Media Research at the Ludwigs-Maximilians-University in Munich, Germany. After his dissertation and habilitation at the Catholic University of Eichstätt he had a visiting professorship at the University of Leipzig (2001/02) and a professorship at the University of Münster (2002–2011). His fields of interest include the internet (public sphere, online journalism, activities of the press and broadcasting media on the internet, search engines, social media), journalism theory and media quality. His recent publications include *Grundlagen des Journalismus* (with Peter Kapern; Springer, 2013) and *An Investigation of Influentials and the Role of Sentiment in Political Communication on Twitter During Election Periods* (with Linh Dang-Xuan, Stefan Stieglitz and Jennifer Wladarsch; in: Information, Communication & Society 16:5, 2013, 795–825).

John V. Pavlik is Professor in the Journalism and Media Studies Department at the School of Communication and Information at Rutgers, the State University of New Jersey, where he has been on the faculty since 2002. While on leave from Rutgers in 2013, he served as the inaugural Associate Dean for Research at Northwestern University in Qatar. He has written widely on the impact of new technology on journalism, media and society. He recently completed a two-year project funded by UNESCO on curriculum reform in journalism and mass communication in Iraqi higher education. His books include *Converging Media: A New Introduction to Mass Communication* (with Shawn McIntosh; Oxford University Press, 2003, 2010, 2011, 2013); *Media in the Digital Age* (Columbia University Press, 2008) and *New Media Tech-*

nology: Cultural and Commercial Perspectives (Allyn & Bacon, 1996, 1998, 2000).

Martin Schreiber was Researcher at the Chair of Cultural and Media History at Saarland University, Saarbrücken, until March 2014 and is now Official Advisor for Quality Management at Trier University. He is currently completing his Ph.D on *The Technologization of Journalism in the Digital Age: Reconfigurations of Journalistic Work by the Usage of Digital Computer Technology since the 1970s*. His research interests are in media history, media technology and journalism, the history of knowledge organization/management and computational methods in history. His publications include *Graphbasiertes Record Linkage. Grundlagen, Methoden und Anwendungen im Bereich bibliografischer Daten* (VDM, 2008), *Internetnutzung in der ländlichen Gesellschaft* (in: Zeitschrift für Agrargeschichte und Agrarsoziologie, 58:2, 2010, 65–76) and *Die Stadt im Zeitalter der vernetzten Kommunikation* (in: Clemens Zimmermann, ed., Stadt und Medien. Vom Mittelalter bis zur Gegenwart. Böhlau, 2012, 265–285).

Mitchell Stephens is a Professor of Journalism at New York University's Carter Institute. He has reported on globalization for the public radio program *Marketplace* and he has been history consultant to the *Newseum*. He is the author of *A History of News* (Oxford University Press, 1988, 1997, 2007), which was a New York Times "notable book of the year" and has been translated into five languages, and *The Rise of the Image, the Fall of the Word*. He published three books in 2014: *Imagine There's No Heaven: How Atheism Helped Create the Modern World* (Palgrave-Macmillan), *Journalism Unbound: New Approaches to Writing and Reporting* (Oxford University Press) and *Beyond News: The Future of Journalism* (Columbia Journalism Review Books), from which the excerpt included here is taken.

Karin Wahl-Jorgensen is Professor in the Cardiff School of Journalism, Media and Cultural Studies at Cardiff University. Her research focuses are in the relationship between journalistic practice and citizenship. She is the author or editor of five books, most recently *Disasters and the Media* (with Simon Cottle and Mervi Pantti; Peter Lang, 2012) and the *Handbook of Journalism Studies* (co-edited with Thomas Hanitzsch; Routledge, 2009). She has also published more than 40 journal articles and 20 book chapters. She is currently finalizing *Emotions, Media and Politics* (Polity Press, forthcoming

2015), and has just completed a large-scale content analysis for the BBC Trust on the breadth of opinion coverage in British public service broadcasting. She is Associate Editor of *Communication Theory* and she is a frequent media commentator on topics relating to citizenship and the media.

Jürgen Wilke, Ph.D. 1974, habilitation thesis 1983, 1984–1988 Professor and Chair for Journalism at the Catholic University Eichstätt, 1988–2012 Professor for Communication Research at the Johannes Gutenberg-University Mainz. Visiting Scholar at the University of Washington (Seattle) in 1993 and 1999, Università della Svizzera Italiana (Lugano) 2001 and 2011, Prof. h.c. Lomonossow University (Moscow) since 2004, Corresponding Member of the Austrian Academy of Sciences (Vienna) since 2005. His areas of research include Media Structure and Media History, News and News Agencies, Political Communication and International Communication. His numerous publications include *Grundzüge der Medien- und Kommunikationsgeschichte* (Böhlau, 2000, 2008) and *Fernsehwelten: Auslandsnachrichten im deutschen Fernsehen* (with Thorsten Quandt, Christine Heimprecht and Thilo Pape; Springer, 2014).

Clemens Zimmermann holds the Chair of Cultural and Media History at Saarland University, Saarbrücken. He has published extensively on rural-urban relations, the European metropolis and media history. He is the author of *Die Zeit der Metropolen: Urbanisierung und Großstadtentwicklung* (Fischer, 1996, 2000, 2004, 2012) and *Medien im Nationalsozialismus. Deutschland 1933–1945, Italien 1922–1943, Spanien 1936–1951* (Böhlau, 2007). His edited books include *Industrial Cities: History and Future* (Campus, *2013), Creative Urban Milieus: Historical Perspectives on Culture, Economy, and the* City (with Martina Heßler; Campus, 2008) and *Politischer Journalismus,* Öffentlichkeiten *und Medien im 19. und 20. Jahrhundert* (Thorbecke, 2006). He co-edits *Informationen zur modernen Stadtgeschichte* (IMS), *Zeitschrift für Agrargeschichte und Agrarsoziologie* (ZAA) and *Beiträge zur Stadtgeschichte und Urbanisierungsforschung.*

Index

Cultural Studies

Babette Bärbel Tischleder
The Literary Life of Things
Case Studies in American Fiction
2014. Ca. 300 pages. ISBN 978-3-593-50006-5

Laura Bieger, Christian Lammert (eds.)
Revisiting the Sixties
Interdisciplinary Perspectives
on America´s Longest Decade
2013. 343 pages. ISBN 978-3-593-39990-4

Clemens Zimmermann (ed.)
Industrial Cities
History and Future
2013. 368 pages. ISBN 978-3-593-39914-0

Susanne Hamscha
The Fiction of America
Performance and the Cultural
Imaginary in Literature and Film
2013. 334 pages. ISBN 978-3-593-39872-3

Gertraud Koch,
Stefanie Everke Buchanan (eds.)
Pathways to Empathy
New Studies on Commodification,
Emotional Labor, and Time Binds
2013. 213 pages. ISBN 978-3-593-39894-5

Nancy Konvalinka
Gender, Work and Property
An Ethnographic Study of Value
in a Spanish Village
2013. 294 pages. ISBN 978-3-593-39661-3

Christian Huck,
Stefan Bauernschmidt (eds.)
Travelling Goods, Travelling Moods
Varieties of Cultural Appropriation
(1850–1950)
2012. 261 pages. ISBN 978-3-593-39762-7

Hans Peter Hahn, Karlheinz Cless,
Jens Soentgen (eds.)
People at the Well
Kinds, Usages and Meanings of
Water in a Global Perspective
2012. 316 pages. ISBN 978-3-593-39610-1

Michael Nentwich, René König
Cyberscience 2.0
Research in the Age of Digital
Social Networks
2012. 237 pages. ISBN 978-3-593-39518-0

Monika Grubbauer,
Joanna Kusiak (eds.)
Chasing Warsaw
Socio-Material Dynamics of
Urban Change since 1990
2012. 336 pages. ISBN 978-3-593-39778-8

Michi Knecht, Maren Klotz,
Stefan Beck (eds.)
Reproductive Technologies as Global Form
Ethnographies of Knowledge, Practices,
and Transnational Encounters
2012. 386 pages. ISBN 978-3-593-39100-7